Smart Buildings, Smart Communities and Demand Response

Engineering, Energy and Architecture Set

coordinated by
Lazaros E Mavromatidis

Volume 8

Smart Buildings, Smart Communities and Demand Response

Edited by

Denia Kolokotsa
Nikos Kampelis

ISTE

WILEY

First published 2020 in Great Britain and the United States by ISTE Ltd and John Wiley & Sons, Inc.

ISTE Ltd
27-37 St George's Road
London SW19 4EU
UK

www.iste.co.uk

John Wiley & Sons, Inc.
111 River Street
Hoboken, NJ 07030
USA

www.wiley.com

Library of Congress Control Number: 2020945010

British Library Cataloguing-in-Publication Data
A CIP record for this book is available from the British Library
ISBN 978-1-78630-426-1

Contents

Preface

Background

Demand response (DR) is associated with significant environmental and economic benefits when looking at how the electricity grid can operate optimally. Adding flexibility in power consumption provides a sound basis for improving the grid's environmental performance and efficiency. For example, reducing peak loads at grid level could lead to a lower level of operation for generation plants with a high running cost, low efficiency and low environmental performance. Furthermore, as the storage of electricity is bound to technical and economic constraints, the absorption of excess electricity from renewable energy sources is feasible through a demand following generation concept.

DR is gradually gaining ground with respect to (1) the reduction of peak loads; (2) grid balancing; and (3) dealing with the volatility of renewable energy sources (RES). In this context, demand side management techniques such as peak clipping, valley filling, load shifting and flexible load shape are already being employed. Also, various DR programs are being designed and implemented, including critical peak pricing, capacity bidding, thermostat/direct load control and fast DR dispatch/ancillary services.

In DR, the consumer becomes a prosumer with an important active role in the exchange of energy on an hourly basis. This transition calls for high environmental awareness and new tools and services which will improve the dynamic, as well as secure multidirectional exchange of energy and data. Overall, DR is identified as an important field for technological and market innovations aligned with climate change mitigation policies and the transition to sustainable smart grids in the foreseeable future.

Why this book?

This book provides an insight into various intrinsic aspects related to the assessment of DR potential, at the building and the community level. Issues pertaining to the use of building energy models, compared to actual performance, and smart monitoring are addressed. Furthermore, temperature set-point adjustment, which is a standard practice in controlling heating, ventilation and air conditioning (HVAC) systems, is assessed with the aid of simulation, to investigate the optimality in multidynamic systems' operation. On the other hand, the book focuses on load shifting optimization at the community level on the basis of time of use (ToU) and real-time pricing (RTP). The rationale behind this is that energy markets should be operated in a transparent manner inducing higher efficiency of power grids through the promotion of renewable energy. In this context, it is foreseen that a high penetration of RES, given their minimal operational expenses and environmental advantages, should be reflected in the time slots of low costs and consumer prices. In this way, all consumers will be provided with a clear roadmap and the necessary motivational factors in order to adjust our consumption when possible and take advantage of electric energy availability from clean resources.

Who is this book for?

This book focuses on near-zero energy buildings (NZEBs), smart communities and microgrids. Therefore, on one hand, it would be valuable for experts, professionals and postgraduates with an interest in (1) highly efficient buildings and communities; (2) smart monitoring systems; and (3) building energy modeling. On the other hand, the book would be beneficial for professionals with an interest in building or community level power predictions and optimization, as well as about how such tools and techniques can be utilized to evaluate DR at the building and/or district level.

Structure

Firstly, a comprehensive approach for evaluating the performance of industrial and residential smart energy buildings/NZEBs is presented. A detailed audit of construction characteristics, installed systems and controls is conducted and presented. Subsequently, holistic data from advanced metering and sensor equipment are explored to verify energy consumption and actual building energy performance. Dynamic energy models are developed, validated and tested to explore key aspects of the operational behavior of buildings and systems, and draw essential knowledge about their performance. Consumption data based on real measurements is compared, on one hand, with dynamic building model simulation results and on the other hand, with the initial annual energy consumption, obtained via the building's energy efficiency certification scheme prior to construction. Findings are explored to address the actual performance gap, reflect on the limitations of each approach and highlight important conclusions.

Secondly, the book focuses on how DR can be applied at the building level. A novel evaluation and optimization methodology, in the context of the building level DR, is presented. To this end, DR is assessed with the aid of an RTP scheme based on the actual energy market data. In this context, HVAC system performance is evaluated according to the energy consumption, the corresponding energy costs and the indoor thermal comfort.

Thirdly, the book describes how DR can be applied at the community level by exploiting predictions of day-ahead consumption and/or production and load shifting. The benefits of this approach are evaluated in terms of the economic savings based on a flat versus ToU tariff and an RTP scheme. The reliable prediction of power consumption and/or production 24 hours ahead is performed using artificial neural network modeling, whereas load shifting optimization is conducted using a genetic algorithm dual-objective optimization algorithm.

In Chapter 2, the smart and zero energy building facilities used as case studies for evaluating DR at the building and the community levels are presented.

Chapter 3 provides a thorough analysis of the performance of residential and industrial buildings with the aid of measurements and how they can be utilized for building energy modeling and validation purposes.

Chapter 4 presents a newly developed approach for optimizing the operation of HVAC systems from a DR perspective.

Chapter 5 presents a novel approach for the community level prediction and optimization in a DR setting.

Finally, the overall conclusions and recommendations arising from the findings of this research are presented.

Acknowledgments

The editors express their deepest appreciation to all the authors for their contribution and to the European Commission, for allocating the funds in order for the Smart GEMS project to be implemented. Special thanks are owed to Dr. Cristina Cristalli, Head of Research for Innovation in the Loccioni Group and to the Loccioni Group for providing access and support for research activities in the framework of Smart GEMS project to be conducted in their industrial high-end facilities.

Nikos KAMPELIS
September 2020

Nomenclature

Acronyms

AC	Alternating Current
AMI	Advanced Metering Infrastructure
ANN	Artificial Neural Network
ARC	Aggregators or Retail Customers
AS	Ancillary Services
BEMS	Building Energy Management System
biPV	Building-Integrated PhotoVoltaic
CHP	Cogeneration of Heat and Power
$CO_{2\text{-eq}}$	Carbon Dioxide Equivalent Emissions
COP	Coefficient Of Performance
CPP	Critical Peak Pricing
CSP	Curtailment Service Provider

Chapter written by Nikos KAMPELIS.

Cv	Coefficient of Variance
DA	Day Ahead
DARTP	Day-Ahead Real-Time Pricing
DC	Direct Current
DEMS	District Energy Management Systems
DER	Distributed Energy Resources
DG	Diesel Generator
DHW	Domestic Hot Water
DR	Demand Response
DRP	Demand Response Providers
DSM	Demand Side Management
DSO	Distribution System Operator
EED	Energy Efficiency Directive
EER	Energy Efficiency Ratio
EMS	Energy Management System
ESCO	Energy Service COmpany
FC	Fuel Cell
GA	Genetic Algorithm
HRES	Hybrid Renewable Energy System
HVAC	Heating, Ventilation and Air Conditioning

ID	Integrated Design
IoT	Internet of Things
IPMVP	International Performance Measurement and Verification Protocol
MAPE	Mean Absolute Percentage Error
MBE	Mean Bias Error
MILP	Mixed Integer Linear Programming
MINLP	Mixed Integer NonLinear Programming
MIP	Mixed Integer Programming
MPPT	Maximum Power Point Tracking
MT	Micro-Turbine
NARX	Nonlinear AutoRegressive ANN with eXogenous input
NIST	National Institute of Standards and Technology
NZEB	Near-Zero Energy Building
OpenADR	Open Automated Demand Response
PMV	Predicted Mean Vote
PPD	Percentage of People Dissatisfied
PSO	Particle Swarm Optimization
PV	PhotoVoltaic

RES	Renewable Energy Sources
RH	Relative Humidity
RMSE	Root Mean Squared Error
RTO	Regional Transmission Operator
RTP	Real-Time Pricing
SaaS	Software as a Service
SDG	Sustainable Development Goal
ToU	Time of Use
VEN	Virtual End Node
VTN	Virtual Transfer Node
WT	Wind Turbine
ZEB	Zero Energy Building

Symbols

C_i	Day-ahead price per hour for hours 1–24
$C\_(E,T)$	Total energy plus taxes (€)
$C_{E\_unit}^{h}$	Day-ahead hourly unit cost of energy in each building (€/kWh)
C_T	Total tax charges (€)
C_S	Energy procurement cost (€)
C_N	Network services cost (€)

$C_{S,F}$	Energy procurement fixed cost component (€/kWh)
C_{EDD}	Daily excise duty on electricity and taxes (€)
$C_{v,u}$	Various costs normalized per kWh (€/Wh)
C_F	Fixed cost component (€)
C_{Pmax}	Maximum power cost component (€/kW)
C_{AT}	Active energy cost component (€/kWh)
C_{A-UC}	Fixed cost for up to 4 GWh per month (€/kWh)
C_{EDH}	Excise duty per kWh (€/kWh)
C_{FAA}	Parameter to account for F, AT, and A-UC components (€/kWh)
$C_{Pmax,F}$	Maximum power fixed cost component (€/kW)
Icl	Clothing insulation (m^2K/W)
IVA	Value added tax (€)
$Load_{Shift}$	Daily load shift (kWh)
$X^h_{E_{opt}}$	GA optimized hourly electrical energy (kWh) at building or building group level
M	Metabolic rate (W/m^2)
P_i	Hourly average power consumption of the HVAC in kW (equivalent to kWh)
$T_{s_{i=1}}^{24}$	Hourly temperature set points of the HVAC system the next day
$Cost_E$	Daily energy operating costs (€)

$Cost_{E\_Lab}$	Daily energy operating costs of Leaf Lab (L4) building (€)
$Cost_{E\_Summa}$	Daily energy operating costs of Summa (L2) building (€)
$Cost_{E\_Kite}$	Daily energy operating costs of Kite (L5) building (€)
DA_h	Day-ahead market prices (€/kWh)
$DA_{N,h}$	DA price flexible factor per hour h (€/kWh)
R	Pearson's coefficient
RH	Relative humidity (%)
T_{air}	Air temperature (T_{air}) (°C)
Tr	Mean radiant temperature (°C)
V_{air}	Relative air velocity (m/s)
W	Effective mechanical power (W/m^2)
w_c	Weighting coefficient for the daily operational cost of energy for the HVAC
w_{pmv}	Weighting coefficient for the daily thermal comfort
X_E^h	Hourly value of total energy consumption in each building (kWh)
$X_{E_{baseline}}^h$	Baseline hourly electrical energy (kWh) based on day-ahead neural network predictions

1

Demand Response in Smart Zero Energy Buildings and Grids

1.1. Introduction

In broad terms, demand response (DR) refers to retail customers participating in electricity markets by responding to varying prices over time (U.S. Department of Energy 2006). DR is otherwise defined as "changes in electric usage by end-use customers from their normal consumption patterns in response to changes in the price of electricity over time, or to incentive payments designed to induce lower electricity use at times of high wholesale market prices or when system reliability is jeopardized" (Losi *et al.* 2015). However, DR is inextricably linked to smart grids since an optimum response to real-time signals or any kind of dynamic information requires interoperability, embedded intelligence and advanced controls working harmonically in the same direction.

In parallel, energy efficiency in the building sector calls for innovations, effective policies and regulations to enable a new technological paradigm for new and renovated dwellings. In particular, the design and construction of smart energy

---

Chapter written by Nikos KAMPELIS.

building and zero energy building (ZEB) as well as smart communities is a primary objective that needs to be met as part of the smart grid evolution. In this context, aggregating buildings' energy is under certain conditions a valuable resource allowing indirect participation in energy markets.

Smart communities can be formed in various physical configurations. Connecting building facilities together to form semi-independent small subdivisions of the main electrical distribution grid offers significant advantages through exploiting on-site energy generation by applying advanced control, intelligence and storage at the local level. The microgrid paradigm fits well with the smart community concept especially with respect to the basic local interconnecting infrastructure and the overall management of energy consumption, renewable energy production and storage. Specifically, the microgrid concept is bound to the necessary operations for monitoring, storing and controlling energy flows between smart buildings and other facilities, i.e. storage so that renewable energy is optimally deployed.

In this framework, the state of the art in the fields of smart energy buildings and ZEBs is presented first. Recent advances in DR at the building and district level are explored with a specific focus of DR in microgrids.

1.2. Smart and zero energy buildings

The smart building is a fundamental entity of the smart grid concept. Nonetheless, providing demand flexibility and operational responsiveness in a smart building remains a challenge because it requires a high level of intelligence along with the integration and optimization of users' actions and decisions. The smart building combines advanced energy management systems (EMSs) overseeing the operation of a range of elegant and multifunctional intelligent equipment to control various building systems such as heating, ventilation and air conditioning (HVAC), lighting and shading. The smart building user is informed of the building's energy flows and provided with the tools for the dynamic management of systems installed, e.g. to adjust indoor environment conditions according to his/her preferences or specific needs, control devices remotely, etc. Furthermore, tools assisting users to optimize the energy performance of a smart building and at the same time minimize the cost of the energy bills are envisioned.

In this context, the goal of the efficient exchange of energy and information between the building and the grid in a way that is mutually beneficial must be facilitated. At the distribution level, the energy demand in buildings forms an important asset in terms of the collective power flexibility potential.

Aligned with the smart building, the (near-)zero energy building (NZEB) concept constitutes a technological paradigm of unquestionable importance because it incorporates the necessary measures for minimizing the net energy inflow from the main grid. The NZEB is inherently associated with integrated design (ID), high-end energy conservation measures, advanced controls as well as on-site renewable energy generation and exploitation. The NZEB concept resembles the evolution of building design and construction in a holistic way to ensure the true and actual sustainable levels of energy performance.

The concepts of smart energy buildings and ZEBs have attracted the interest of the scientific community, policy organizations and the industry worldwide. Special attention is paid to coupling ID, energy efficiency and renewable energy in new and renovated buildings. From a policy perspective, this is being pursued via strategic energy and environmental objectives, policy initiatives, regulatory reforms and financial incentives. In this regard, the EU has emphasized on the reduction of the high energy consumption in the building sector using various policy tools and directives including, among others, the EU 2020 targets, the Energy Performance Building Directive (EPBD), the climate change adaptation and mitigation strategies and the low carbon economy roadmap 2050 (European Parliament and the Council of the European Union 2010; European Commission 2011).

EPBD recast (European Parliament and the Council of the European Union 2010) imposed member states (MS) to ensure all public buildings (or buildings used by public organizations) as well as new buildings comply with near-zero energy consumption since 2018 and by 2020, respectively. Under this legislative framework, MSs are responsible for reporting on the detailed progress with respect to NZEBs' agenda implementation in practice, as it needs to be adjusted to reflect national, regional or local conditions.

The NZEB is conceptualized in the EPBD and characterized by a very high energy performance, a very low amount of required energy and a very significant contribution of renewable energy sources (RES) to cover the remaining energy use. Very high energy performance is translated into buildings integrating passive and active systems and falling into the top categories of the energy certification process.

A clear universal definition of a ZEB is, however, somewhat of a challenge and usually linked to the framework of the analysis, i.e. whether carried out for new construction, energy efficiency evaluation or classification, specific research, development of policy tools or another purpose. Definitions may vary according to the metric and period of balance, type of energy use and balance, renewable supply options, connectivity with the grid, requirements, etc. (Marszal *et al.* 2011). Apart from the EPBD, linking energy performance to annual normalized primary energy consumption (kWh/m^2/year), various definitions have been proposed, including net-zero site energy, net-zero source energy, net-zero energy cost and net-zero energy emissions depending on the metric (energy, cost, $CO_{2\text{-eq}}$ emissions) and domain (site or source). Where applicable, a net-zero site energy benchmark is considered most appropriate as it is fully verifiable through on-site measurements and cannot be affected by external factors (i.e. related to the operation of the main grid or the energy market), which may vary according to the dimensions of time, space and territory.

It is noteworthy that quantitative targets linked to zero or near-zero energy performance are dispersed between 0 and 270 kWh/m^2/year of primary energy consumption. Higher figures in this range are associated with hospitals or non-residential buildings (D'agostino *et al.* 2016). For NZE residential buildings, the average targets vary from 33 kWh/m^2/year in Croatia and 45–50 kWh/m^2/year for the many EU MSs (Belgium, Estonia, France and Ireland), while some countries use non-dimensional values or an energy performance class (e.g. A++ in Lithuania) (Groezinger Jan *et al.* 2014). In Italy, the regulation for new dwellings requires a minimum energy efficiency of 65–70 kWh/m^2 (Zangheri *et al.* 2012). In Cyprus, the threshold for NZEB is 100 kWh/m^2/year of primary energy for new and existing residential buildings and 125 kWh/m^2/year of primary energy for non-residential buildings (D'agostino *et al.* 2016).

ZEB or NZEB currently in operation, or even those in development stages, primarily use fossil fuel-based energy sources coupled with renewables such as solar, wind, geothermal or biomass to attain "nearly zero energy" behavior (Hemerlink Andreas *et al.* 2013; Groezinger Jan *et al.* 2014). The transition to smart ZEBs from an industrial point of view depends to some extent on the adoption of common communication protocols, standards and interfaces to enable interoperability of systems, subsystems and the bidirectional flow of energy and information (Kolokotsa *et al.* 2016). Coupling existing building energy systems with modern monitoring and control equipment is often a barrier for renovating the existing building infrastructure.

Discussions in this direction expand toward the challenges of NZEB integration in smart grids with the aid of evolving technologies (Kolokotsa *et al.* 2011; Karlessi *et al.* 2016). Various efforts have dealt with optimizing the design and operation of building integrated renewables, thermal or electrical storage and holistic energy management using a broad range of techniques. Attention has also been drawn to the development of tools for user/customer engagement, increasing transparency of grid operations with the aid of advanced metering infrastructure (AMI) and enabling DR within the Internet of Things (IoT) applications (SEDC 2015).

In many cases, the actual operating performance of buildings significantly deviates from the designed target. This "performance gap" is associated with (1) the design and construction processes of the building envelope and systems otherwise referred to as the "design and construction phase"; or (2) the management of the building and its facilities or the "operational phase". In the design and construction phase, the performance gap is often related to the assumptions/inputs or misuse of calculation methodologies and tools utilized. Furthermore, the performance gap may be linked to the lack of consideration or expertise about the deployment of ID principles impacting energy consumption, indoor comfort and health conditions. Performance gap issues are also evident during the construction phase due to improper installation of building envelope components (i.e. insulation and glazing), which may be a result of inadequate training, time limitations, cost-cutting constraints or barriers related to a resistance to change (Menezes *et al.* 2012; De Wilde 2014). Such phenomena may have as a consequence the occurrence of thermal bridges and high infiltration rates, eventually leading to energy losses, high total energy consumption and unhealthy or uncomfortable indoor conditions. Last but not least, energy management and operational inefficiencies are critical to the observed gap in buildings' energy performance, depending on the specificities of each case. This may be due to a lack of appropriate maintenance and service, misuse of energy systems' operation or suboptimum performance in systems' integration. However, often there is a valid potential for bridging the "gap" of underperformance in the buildings' operational phase, which can be effectively addressed via a mixture of technological, organizational and training actions.

In terms of the technological progress, indoor environmental quality control and building energy management systems (BEMSs) have evolved considerably over the last decades, in parallel with the growing concern about energy efficiency requirements and the demand for environment-friendly buildings. Modern customized building energy management solutions can be exploited to enable better visual, thermal comfort and air quality control. Research efforts in this direction focus on advanced BEMSs, which can implement sophisticated algorithms capable of predicting and evaluating a range of alternatives in the way buildings exchange energy with the ambient environment and the grid. State-of-the-art BEMS techniques nowadays offer the potential for applying predictive control, which may contribute to 20–30% in the reduction of energy consumption (Kolokotsa *et al.* 2009a, 2009b; Papantoniou *et al.* 2015) and equivalent operational cost savings. Prediction of energy demand is becoming increasingly effective as part of an overall energy management optimization process, which could be deployed in the near future (Kolokotsa *et al.* 2009a; Papantoniou *et al.* 2015). Simultaneously, researchers are providing new scientific evidence on how the prediction of renewable energy production can increase its utilizability in building integrated applications and deal with the volatility of distributed energy resources (DER) and the future microgrids.

Furthermore, smart metering, data processing and interpretation provide useful steps for a deeper understanding of buildings energy behavior (Kolokotsa 2016). This is especially important when such knowledge can be developed to inform decisions about the systems' operational strategies based on scientifically sound methodologies and technologically robust processes. In this direction, DR techniques have been applied in various settings to optimize the operation of building energy systems (i.e. HVAC), to perform active load management and to minimize energy from the grid as well as the respective costs on the demand side (Gao *et al.* 2015; Gao and Sun 2016). Accordingly, data monitoring, i.e. the provision of meaningful information along with practical tools for managing energy consumption, in combination with specific incentives, provides the fundamentals for actively engaging users in realizing the potential of DR wide-scale environmental and social benefits. This transition requires targeted investments both in grid infrastructure and at the users' side as well as a transparent, open and attractive regulatory framework to create a supporting environment for innovations, which will transform the market in this field.

With respect to renewables, advanced solutions such as concentrating solar thermal technologies have emerged to offer attractive options in meeting the cooling demand during the summer season and in reducing the heating demand from the grid during the winter season. The real challenge with such systems concerns the design of a suitable and cost-efficient solution utilizing maximum heat from the sun to fulfill the required energy demand (Kalkan *et al.* 2012).

Other commercially available solutions include building-integrated photovoltaics (biPVs) and small wind turbine systems offering a broad range of designs and technical attributes. Such systems are coupled with inverters normally equipped with maximum power point tracking (MPPT) and controls for providing energy to the power grid, microgrid or autonomous installations (Koutroulis *et al.* 2006, 2009; Koutroulis and Kalaitzakis 2006). Recently, building integrated combined solar and wind-driven energy systems have entered the market, promising to be a cost-viable breakthrough technology.

1.3. DR and smart grids

Storage of electricity is subjected to technical and economic barriers making absorption of excess electricity by RES feasible through a demand following generation concept (Kolokotsa *et al.* 2019). DR refers to ways of altering the power consumption of buildings, settlements or other facilities, within a specific time frame, for economic return (Albadi and El-Saadany 2008). It implies regulatory, technological and market changes affecting the way energy is transacted and exploited. DR is strongly interlinked with the smart grid technological paradigm. By definition, in DR, consumers are able to adjust their power purchasing patterns according to the dynamic exchange of information, incentives and disincentives (Siano 2014). In the case of a power system integrating multiple energy carriers, the concept of integrated DR is used along with the energy Internet to provide a wider framework of DR features and complementarity between the various energy sources (Cheng *et al.* 2018).

The smart grid is defined by the National Institute of Standards and Technology of the U.S. Department of Commerce as being composed of seven domains, as shown in Figure 1.1. In the market domain, the trading of grid assets and resources such as electricity, CO_2 allowances, gas and coal takes place, while the operations domain concerns the overseeing of energy management and the smooth control of the power grid transmission and distribution networks by regulating authorities. The service provider domain is linked to the business functions between power system producers, distribution system operators (DSOs) and customers such as billing and customer account management, but also hosts more advanced services supporting energy management and generation. Furthermore, the generation domain is associated with the conversion and supply of electricity from various energy carriers such as gas, coal, pumped hydro, wind, solar and geothermal. New requirements include greenhouse emissions control, an increase of RES and provision of storage to deal with RES intermittency. The transmission domain is dealing with the reliable transfer of electrical power from generation to distribution substations. The distribution domain refers to the link between the customer and the transmission domains, which takes place through various network configurations (radial, looped, meshed). Finally, the customer domain is segmented to differentiate between homes, commercial buildings and industrial facilities. The energy services interface is part of the customer domain for establishing remote communication and applications control.

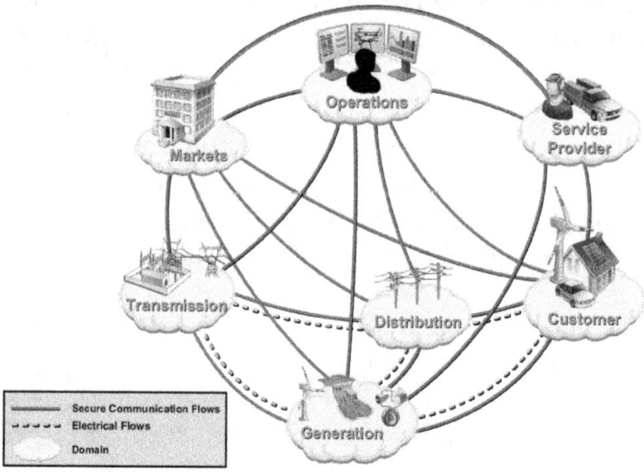

Figure 1.1. *Smart grid NIST conceptual model*

However, DR is linked to sustainable development goal (SDG) 7 for ensuring access to affordable, reliable, sustainable and modern energy for all (Mccollum *et al.* 2017). DR is directly linked to targets for increasing the share of renewables and improving energy efficiency in smart grids. In addition, the wide implementation of DR is expected to be complementary to SDG 13 as part of the efforts to keep global warming well below 2°C above preindustrial levels.

In this context, DER and DR (sometimes the term demand side management (DSM) is used interchangeably) are gradually gaining ground with respect to their potential applications in (1) the reduction of peak loads; (2) grid balancing; and (3) dealing with the volatility of RES. Maintaining grid balance is a primary ancillary service and a prerequisite for the provision of high-quality power utility services affecting everyday life, as well as social and economic progress.

According to Palensky and Dietrich (2011), DSM measures can be categorized based on the timing and the impact of the measure into energy efficiency, ToU, DR and spinning reserve. Another definition of DSM measures is given in Koliou (2016) as a way to induce a load shape change to obtain short- and long-term benefits (Figure 1.2).

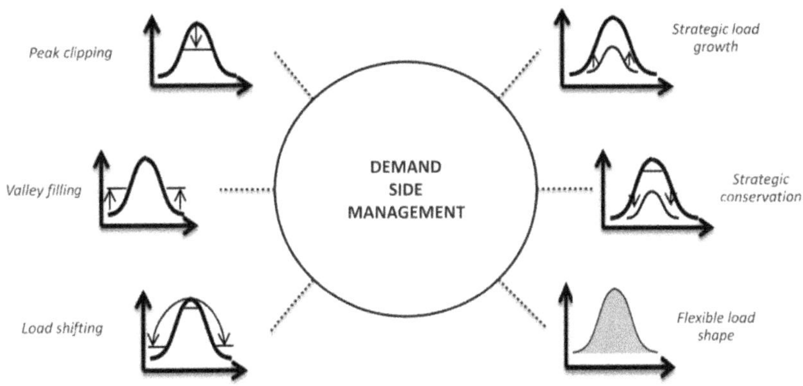

Figure 1.2. *DSM power profile change objectives (Koliou 2016)*

The following list provides a brief explanation of power profile change objectives presented in Figure 1.2 to be pursued in DSM:

– peak clipping refers to the reduction of system peak load using direct load control;

– valley filling concerns the exploitation of energy during low utilization periods to improve the ratio between the peak and minimum load of the system;

– load shifting is related to rewarding end users for time shifting their consumption in order to reduce system peak levels;

– strategic load growth is connected to establishing objectives that will lead to higher electrical energy consumption such as providing tax incentives for e-mobility;

– strategic conservation is associated with total lower energy consumption due to higher overall efficiency;

– flexible load shape is linked to the activation of loads' flexibility in real time to optimize demand and supply.

DR is otherwise classified into (1) incentive-based, including direct load control, interruptible/curtailable rates, emergency DR, capacity market programs and demand bidding programs; and (2) time-based, such as ToU rates, critical peak pricing (CPP) and real-time pricing (RTP). In CPP, a high rate is imposed on the customer in the case of critical events of high wholesale market prices (Song *et al.* 2019). In RTP, end customers are forwarded wholesale market prices a day or 1 h before energy delivery. One of the main challenges in RTP is associated with the requirement for robust and continuous real-time communication between the energy provider and customers (Doostizadeh and Ghasemi 2012). Prices in RTP fluctuate as a consequence of wholesale market price variation and design aspects. Several

RTP structures were assessed by utilities (Doostizadeh and Ghasemi 2012). Other pricing structures such as inclined block rate, whereby tariffs vary based on consumption level thresholds, have been exploited in order to promote energy conservation, load balancing and reduction of peak load (Guang *et al*. 2019).

In this framework, the idea of an open and transparent smart grid accommodating participants on a fair and inclusive basis is tied with (1) the allocation of actual costs for the generation, transmission and distribution to the various stakeholders; and (2) the transition to a very high share of clean energy resources in the electricity generation mix. Undoubtedly, the smart grid of the next decade needs to ensure very high penetration of RES, as well as gradual replacement of fossil fuels and other power sources associated with high environmental risks. Grid energy efficiency is currently related, among others, with requirements for significant levels of spinning reserves and low-efficiency generators compromising environmental sustainability. In DR, consumers are incentivized to control their consumption in time to follow high RES production, contribute to the decrease in demand peaks and lead to improved overall energy efficiency at the grid level.

The potential benefits of DR for customers, system operation, market efficiency and reliability of the power system were critically evaluated based on different innovative technologies, real case studies and research projects (Siano 2014; Li *et al*. 2018, 2019). The long-term impact of DR in the Portuguese electric system is investigated in Anjo *et al*. (2018). In all of the scenarios studied, DR was found to lead in reduction of the total costs and of the total capacity as well as an increase in RES penetration. Also, high variable RES power generation is reflected on changes in models dealing with optimization of the power system (Deng and Lv 2019). In this context, short-term operations become increasingly important with respect to integrating renewables, power generation flexibility, interregional transmission of energy, energy storage and DR.

On the contrary, several factors are slowing down the widespread implementation of DR such as human intrinsic, economic, social, technological, and regulatory aspects as discussed in Good *et al.* (2017), Vallés *et al.* (2016) and Torriti *et al.* (2010). In terms of the infrastructure modernization necessary for DR to take place, smart meters and AMI have a fundamental role to play. Advanced metering infrastructure, such as smart meters, is an essential part of the smart grid for utilities to be able to monitor and control any point of energy consumption/production or distribution in the grid. AMI and smart meters are also essential for consumers to be able to monitor their consumption and react to information about prices or DR events in real time. Moreover, AMI constitutes the necessary infrastructure for the collection of load profiles, which can be exploited by utilities or aggregators using clustering to identify common patterns of energy consumption, design appropriate tariffs and target groups of customers for participating in demand response providers (DRP) (Satre-Meloy *et al.* 2020).

Various forms of possible DR program types and interactions between stakeholders involved such as utilities, aggregators and resources are defined in the Open ADR standard (OpenADR Alliance 2014). Specifically, DR program types of CPP, capacity bidding, thermostat/direct load control, fast DR dispatch/ancillary services (AS), electric vehicles and DER are defined. According to the OpenADR 2.0 (Open Automated Demand Response), a resource for a utility in a DR program can be anything from a single customer load or an aggregator down to something as specific as a thermostat.

In Figure 1.3, two simple DR deployment scenarios are presented. When a resource is enrolled in a DR program, the utility may dispatch an “EiEvent” message to the virtual end node (VEN), serving as a means of communication for the resource, specifying the resource to be targeted. If such a target qualifier is not included, then all resources behind a VEN are specified. In this case, the relationship between the DR Program Party and the Resource Party is direct. The Direct 1 scenario applies to commercial and industrial buildings with a VEN gateway translating the incoming signal to a load controller based on a specific protocol. In the Direct 2 scenario, the VEN is part of a Building Management System (BMS) and the resource is composed of large building facilities such as HVAC, lighting and industrial processes.

In Figure 1.4, two more complicated scenarios, employing a facilitator (top) and an aggregator (bottom), are presented. The facilitator is an intermediary assisting asset parties in managing their resources. Resources are directly enrolled in DR programs and remain in direct communication with the DR Program Party. The VEN, in this case, sometimes offered as a cloud-based software as a service (SaaS), resides within the facilitator acting as a gateway for OpenADR actions. When a DR signal is sent by the DR Program Party (VTN) to the facilitator (VEN), it is forwarded to a specific resource for some DR logic to take place or converted to load control commands for several load controllers. A company managing the facilities of a large commercial or industrial company, energy service companies (ESCOs) and cloud-based equipment management, i.e. smart thermostat services, are different example applications that fall in this category. In the second scenario in Figure 1.4, the resources are not directly engaged with the DR Program Party but with the aggregator instead. The aggregator enrolls resources forming various portfolios managed in response to DR signals received by the DR Program Party. The DR Program Party has no knowledge of the resources the aggregator is managing. Instead, the aggregator is the only point of reference for the resources in this scenario.

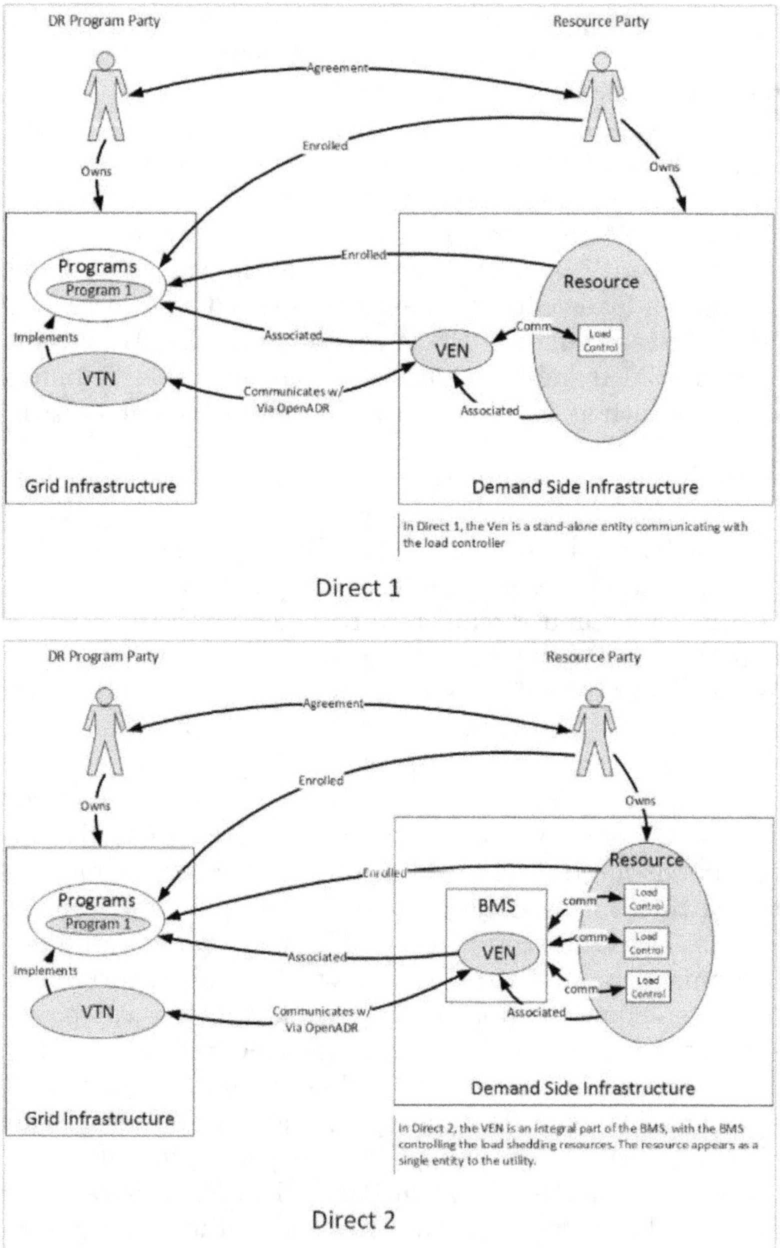

Figure 1.3. *Open ADR 2.0 simple DR deployment scenario (Direct 1&2; OpenADR Alliance 2014)*

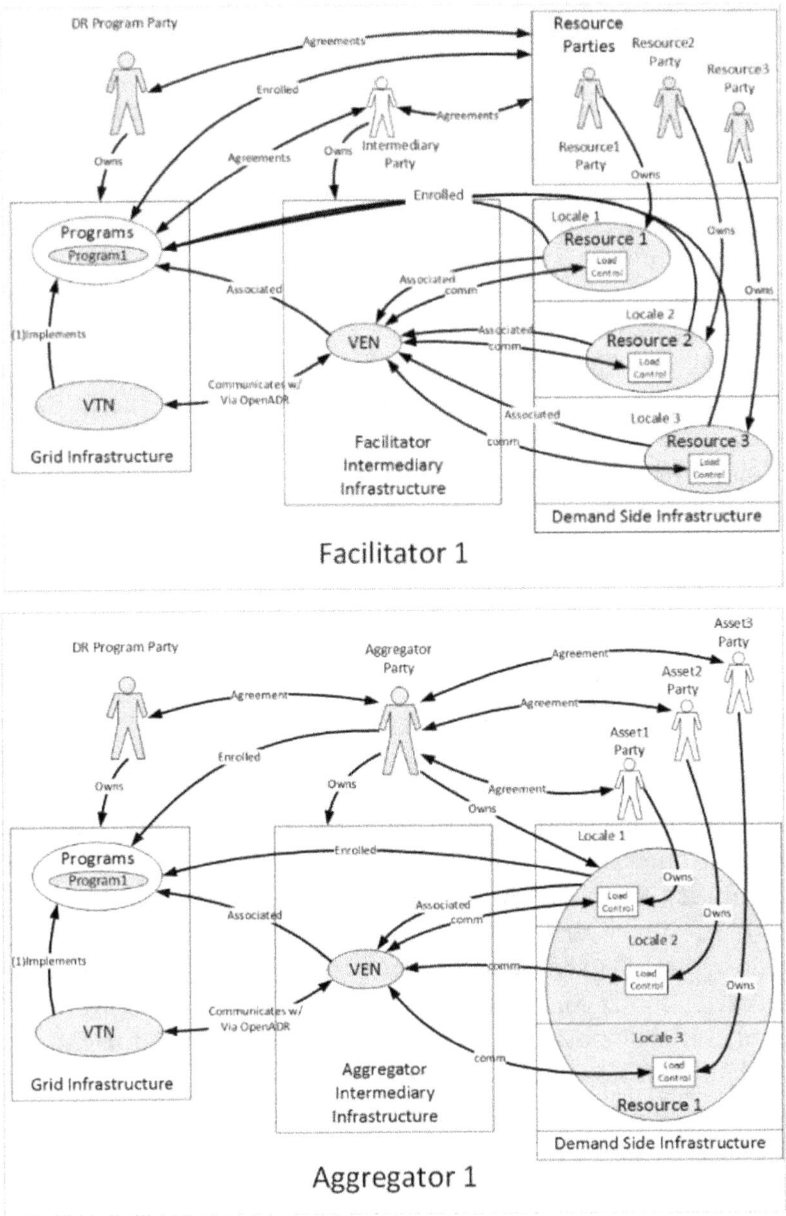

Figure 1.4. *Open ADR 2.0 facilitator and aggregator DR deployment scenarios (Facilitator 1, Aggregator 1; OpenADR Alliance 2014)*

1.3.1. *DR and congestion management*

One of the challenges related to smart grid transition is congestion management. The high penetration of DER in the distribution grids may in some cases cause congestion issues, thus creating the need for new approaches to deal with such a constraint. DER coordination, flexibility and consumer active demand are the basis for the next-generation efficient and reliable distribution grid (Zhang *et al.* 2014a). Dynamic pricing can be exploited in this context to relieve congestion and reduce line losses in distribution networks. Such an approach is proposed in Huang *et al.* (2016) to facilitate the high penetration of electric vehicles. The cost of flexible loads along with the line losses cost due to the network's topology form a single objective function to address the cost of congestion management and yield realistic and optimal results.

Moreover, in Heinrich *et al.* (2020) a local flexibility market trial has been implemented under real conditions and addressed a lot of important issues from the different perspectives of the stakeholders involved. In particular, the authors in Heinrich *et al.* (2020) present a thorough approach for baseline flexibility services and capacity limitation services. In this framework, the DSO creates flexibility requests based on the forecasting of congestion risk in the distribution network. Long-term forecasting in this context uses load scenarios that do not predict the electric load for each node over time but provide a probabilistic network evaluation based on historical data and a decomposition approach. The risk of congestion management is translated into operational cost (i.e. reduced transformer lifecycle and cost of non-provision of services to customers) and compared to the cost of activating flexibility services alleviating this risk. In case the cost of alleviating the risk of congestion is lower than activating flexibility services, the DSO issues a flexibility services list and the maximum price for each of the services on the list. The aggregators may respond to the DSO by sending their bids and ultimately one bid may be accepted to be activated for each flexibility service on the list. The scope of the local flexibility market is to operate in parallel and provide complementary services to the wholesale markets by optimizing the operation of the distribution grid in the presence of increased DERs.

1.3.2. *DR and AS*

Although most balancing in the power system takes place through energy scheduling, real-time contingencies such as the loss of a generator or a major transmission line require a different level of response referred to as AS. AS in the power system include frequency control, voltage control, spinning reserve, standing reserve, operating reserve, black start capability, remote generation control, grid loss compensation and emergency control actions (Yin *et al.* 2016). The value of AS is associated with the capability of the grid to respond in a fast and reliable way and maintain balance. AS requirements vary from 1% of the load in wide interconnected systems to 5–7% in smaller systems with wind and solar generation. DR resources are able to offer significant and in some cases superior AS to the grid (Ma *et al.* 2013). In fact, curtailing loads can be faster than varying the rotational speed of large-scale equipment such as in conventional power generation plants. Furthermore, integrating DR resources in AS opens up the controllable reliability options for system operators and enables greater system flexibility, thus allowing improved penetration of RES such as wind or solar generation.

In particular, activating DR actions in a short amount of time has been identified as suitable in providing contingency and operating reserves. HVAC systems are considered controllable loads that can be exploited in this way since (1) they are installed in most residential and commercial buildings and consume a high share of their electric load; (2) their operation is linked with the building's thermal inertia, therefore allowing a margin for operational control within a range of set points, which is not directly translated to a deviation from the acceptable thermal comfort levels of occupants; and (3) they are coupled with EMSs. Also, thermostatically controlled loads (TCLs) such as refrigerators and water heaters have been considered in studies investigating the potential of DR in connection with AS. In this context, the DR control signal entails a request for resetting the temperature set point of a TCL, as an action related to frequency regulation or to a change in power consumption. In organized wholesale markets, transmission providers procure AS via cost-based contracts and AS costs are defined through a regulated process to include a bid and an opportunity component. AS providers are compensated for their marginal costs (including maintenance and operational costs, i.e. fuel) on the basis of the bid component and in case of energy sale a lost opportunity cost based on the difference between the market clearing price and the AS provider energy market bid. In this context, DR resources can be exploited to provide AS at a different frequency and duration compared to current DR deployment experience. For instance, in the case of emergency and economic load shedding, DR is triggered a limited number of times in a year (10–15) for a duration of 4–8 h each time. Instead, AS such as contingency reserve in particular are deployed at a frequency between once every 2 days and once every 2 weeks for a duration of up to 30 min each time.

Demand fluctuation is a significant cost factor driving ancillary costs to higher levels. This is expected to become more significant as RES penetration rises and poses certain challenges for the conventional power generation units. In Tsitsiklis and Xu (2015), the ancillary costs are modeled with respect to demand variability. A dynamic pricing mechanism is proposed, which motivates customers to adjust their energy consumption while they provide a balanced DR. Results through a dynamic game theoretic approach indicate that demand variability and requirements for peaking plants can ultimately both be significantly reduced.

1.3.3. *DR programs*

Adding flexibility in power consumption provides a sound basis for improving the grid's environmental performance. Reduction of peak loads at the grid level could lead to a lower level of operation for generation plants of high running cost, low efficiency, and low environmental performance. DR potential in the United States alone could lead to peak load reductions of 150 GW, an equivalent of approximately 2,000 peaking plants (Faruqui *et al.* 2009).

A thorough review of existing DR programs available to U.S. commercial and residential customers by numerous independent system operators (ISOs) and regional transmission organizations (RTOs) was provided in Shariatzadeh *et al.* (2015). Such programs include real-time demand, real-time price (RT-Price), day-ahead load response (DALRP), day-ahead demand response program (DADRP) and so on. In RT-Price programs, consumers can choose to reduce their load in real time as a response to prices exceeding a certain value. A detailed classification and survey of DR programs in smart grids with respect to pricing and optimization algorithms is available in Vardakas *et al.* (2015). In day-ahead real-time pricing (DARTP) programs, the predicted next day's RT-Prices are announced to customers and used for billing their consumption. DARTP is reported as one of the solutions, which was tested and found to be effective in achieving flatter demand, a reduction of losses, shorter peak-to-peak distance and a higher load factor.

1.3.4. *Building level DR*

In DR, the consumer becomes a prosumer with an important, active role in the transaction of energy on a day-to-day basis. This transition calls for higher environmental and social awareness as well as new tools and services to allow for a dynamic interoperable bidirectional flow of data. Hence, DR is identified as an important field for technological and market innovations aligned with climate change mitigation policies and the transition to sustainable smart grids in the near future.

In this direction, a wide range of DR techniques was applied and documented according to the type of the loads and the installed facility equipment (Motegi *et al.* 2007; Mathieu 2012). Customers can change their electricity pattern and participate in DR by reducing their use of electricity, by shifting consumption to another time period and by self-generating electricity (Goldman *et al.* 2010). In this context, at the building level, the adjustment of the HVAC temperature set points is reported as an effective way to exploit the thermal mass of the building in order to reduce peaks or shift loads. Changing thermostat settings or reducing the operational time of HVAC systems as well as decreasing artificial lighting levels are some of the main load curtailment techniques (Motegi *et al.* 2007; Goldman *et al.* 2010).

HVAC is among the major energy loads of the building sector (Pérez-Lombard *et al.* 2008). The performance of the HVAC system is of great importance with respect to the energy efficiency of a building overall. HVAC efficiency depends on the technical attributes of the technology employed and on the way systems are controlled, i.e. settings and embedded intelligence, which in turn define its actual operational performance and indoor comfort conditions. Many researchers investigated the potential of applying advanced controls and optimization techniques to improve energy HVAC efficiency (Wang and Ma 2008; Dounis and Caraiscos 2009; Lin and Hong 2013; Afram and Janabi-Sharifi 2014; Papantoniou *et al.* 2015; Harish and Kumar 2016; Okochi and Yao 2016; Afroz *et al.* 2018). In De Rosa *et al.* (2018), the case of CHP (cogeneration of heat and power) combined with thermal and electrical storage is explored in an RTP DR setting for a 12-storey large office building equipped with two 1,300 kW water-cooled chillers and a gas boiler for cooling and heating (including domestic hot water (DHW)), respectively. Savings of 7% are established due to the operation of the thermal storage. Despite the fact that the RTP scheme can be exploited using the Electrical Energy Storage (EES) to store energy during off-peak periods and reduce consumption during peak periods, the high investment cost associated with it is a barrier for its adoption. Furthermore, a multiobjective optimization approach is utilized to investigate the trade-off between the aggregator and smart apartment residents offered RTP in Japan (Susowake *et al.* 2019). A total of 100 smart apartments equipped with photovoltaics (PVs) and batteries used for storage, heat pumps (flexible loads) assisted by solar thermal collectors and electric vehicles (EVs) are used as an aggregator's portfolio for indirect optimization through the identification of RTP profiles to promote load curtailment and load shifting. Optimum results of leveled profits for the aggregator and the apartment residents were obtained and analyzed.

A mixed-integer linear problem (MILP) for real-time cost optimization of HVAC operation at the building level was proposed by Risbeck *et al.* (2017). This study focuses on the optimization of equipment usage in HVAC commercial systems. In their study, building temperature dynamics were either considered linear and used to estimate cooling or heating loads, or assumed to be available as a forecast of hot and cold water demand. Pompeiro *et al.* (2017) applied dynamic programming and genetic algorithm (GA) optimization to maximize thermal comfort and minimize the HVAC cost with PV production and storage in an experimental facility. Their approach concentrated mainly on the exploitation of energy from PV and storage. The operation of the HVAC was controlled based on indoor temperature measurements and its performance was restricted in low, medium and high levels. A ToU pricing scheme of three tariffs was used in the optimization of a small experimental room. An experimental evaluation of an HVAC system under variable pricing was conducted by Serra *et al.* (2014). A linear model of temperature evolution was developed by correlating past values of temperature with HVAC modules turned on/off at any instance in time. The approach was validated in an experimental facility, demonstrating reduced cost with respect to a normal thermostat in two different ToU pricing schemes. In Alhaider and Fan (2015), a mixed integer programming (MIP) configuration was proposed to optimize HVAC operation based on a comfort/cost trade-off. The approach determined when each one out of a set of many HVAC units was turned on and off based on common goals. Cost reductions of 4.73%, 4.5%, 11.2% and 8.5% in two different scenarios were achieved. In Ma *et al.* (2017), direct load control and set point regulation of aggregated HVAC systems for frequency regulation in smart grids were investigated. A simplified HVAC model was used to evaluate temperature evolution and power consumption. Results demonstrated acceptable variations of temperature and on/off operations associated with smaller tracking errors compared to direct load controls and sliding-mode control strategies. In Lee *et al.* (2015), a model predictive control framework was proposed, to determine optimal control profiles of HVAC systems in a DR context. This approach used a nonlinear autoregressive neural network configuration to model the thermal behavior of the building zone and to simulate HVAC control strategies as a response to a DR signal. The optimal control problem was formed as a mixed-integer nonlinear problem (MINLP), taking into account on-site energy storage and energy generation systems with night set-up, demand-limiting and pre-cooling HVAC control strategies. Results for a day in August indicated reliable prediction levels for zone temperature and power. Cost savings, in the case of a varying pricing signal, of 14.25–15.26% for demand-limiting and optimal control without energy generation and storage were achieved. In the case of optimal control combined with energy generation and storage, cost savings of 30.95% were obtained. Particle swarm optimization (PSO) was used in Zhang *et al.* (2014b) to optimize the operation of residential HVAC systems based on a multiobjective scheduling problem, taking into account day-ahead (DA) electricity price, outdoor temperature forecast and user preferences. A cooling scenario with DA pricing was demonstrated where a decrease in HVAC energy consumption of 6.54% and a reduction of 18.71% in electricity cost were achieved.

Furthermore, a bi-level optimization approach is proposed in Nizami *et al.* (2020) with respect to energy management in a residential setting to determine the DA energy quantity bid at the upper level by minimizing energy uncertainty cost while ensuring optimal operation of building loads, DERs and storage. According to the authors, the proposed approach outperforms non-optimal inflexible scheduling methods by up to 51% and deterministic optimization-based methods by 22%.

1.3.5. *District level DR and microgrids*

Hybrid renewable energy systems (HRES) have been implemented in various configurations to combine two or more renewable and non-renewable sources in order to deal with the intermittency of RES, such as solar or wind energy. HRES have important attributes that make them increasingly attractive as alternatives to conventional fossil fuel energy sources in numerous applications (Ou and Hong 2014; Shinkhede and Joshi 2014; Ou *et al.* 2017; Dawoud *et al.* 2018; Guo *et al.* 2018). Microgrid optimal energy management can be highly complex and challenging, especially in the case of hybrid systems combining a wide range of renewable and non-renewable technologies. In Murty and Kumar (2020), optimal dispatch strategy of a hybrid microgrid connecting PV, wind turbine (WT), fuel cell (FC), micro-turbine (MT), diesel generator (DG) and batteries, operating both in standalone and grid-connected operation, is investigated through a multiobjective mixed integer linear programming approach for a particular DR program. Results show a 51.6% reduction of CO_2 emissions in standalone operation. In Yuan *et al.* (2019), stability in microgrids (MGs) in the case of communication interruption is dealt by a hybrid prediction-based DR energy management approach. PSO is used to optimize microgrid operation including a WT, PV, MT, energy storage (ES), FC, interruptible, flexible and fixed loads. Results demonstrate cost reductions of up to 57.89%.

Aligned with HRES, the concept of the microgrid as a semiautonomous system of increased flexibility and manageable energy resources, including renewable energy generation, storage, backup systems and flexible demand, is of particular importance when it comes to supporting grid stability and decentralized control (Hatziargyriou *et al.* 2007). A comprehensive critical review on the EMSs of microgrids, connected to the level of maturity of real-world applications, is conducted by Zia *et al.* (2018).

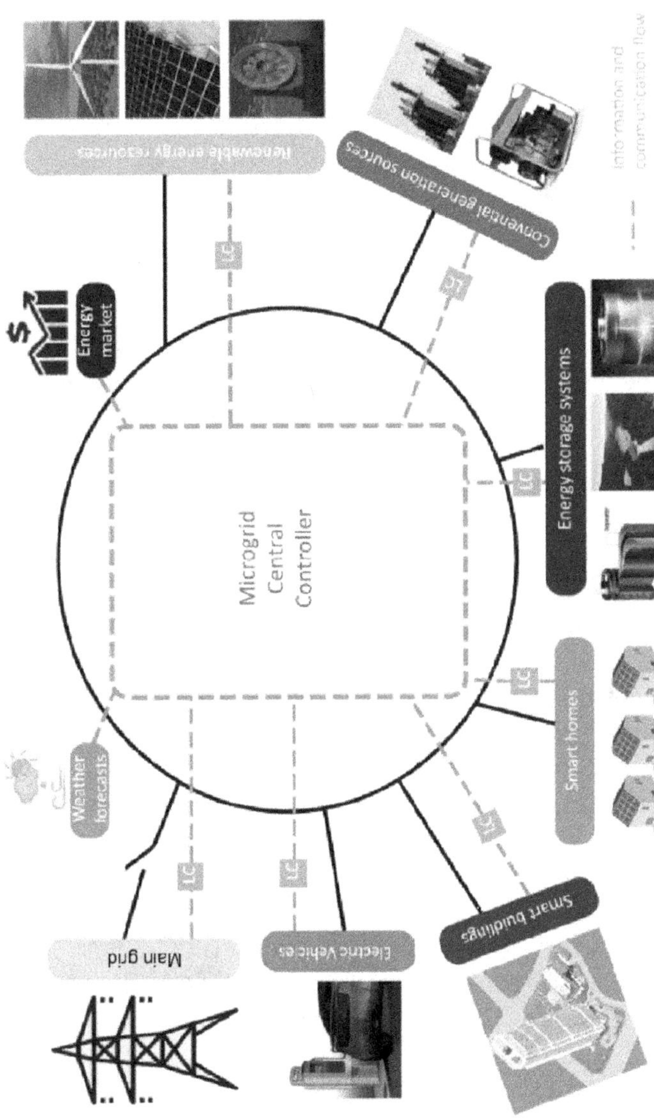

Figure 1.5. *Microgrid conceptual architecture (Zia et al. 2018). For a color version of this figure, see www.iste.co.uk/kolokotsa/smartbuildings.zip*

Communication issues, control technologies and architectures, deployment costs, energy management strategies, optimization, objectives and limitations are addressed. An auto-configuration function using a multiagent approach is proposed by Bui *et al.* (2017) to establish automatic connection or disconnection of DER at the microgrid level, capable of dealing with system faults and reoptimizing the new configuration as necessary. Unsymmetrical and ground faults analysis in microgrids distribution systems is proposed by Ou (2012, 2013). Hirsch *et al.* (2018) surveyed technologies and key drivers of microgrid implementation and research at the international level. Reported drivers in this context include extreme weather-related concerns, cascading outages, cyber and physical attacks, deferral of infrastructure expansion costs, reduced line losses, efficiency improvements, savings, responsiveness, balancing loads and RE generation. In Mohseni *et al.* (2017), the authors present a residential microgrid DA planning approach to accommodate appliance scheduling by modeling, among other aspects, interphase delay duration and time preference, in order to take advantage of shiftable loads and energy storage charging/ discharging time. In Bullich-Massagué *et al.* (2018), multi-microgrid configurations are presented and analyzed by means of the power line technology (AC, DC), layout (series, parallel, mixed) and interconnection technology (transformer, converter). A comparison of architectures based on cost, scalability, protection, reliability, stability, communications and business models is performed. Energy management and DR of multi-microgrids based on a hierarchical multiagent approach by introducing adjustable power are proposed by Bui *et al.* (2018). Different operation modes are evaluated according to a two-level management cooperative multi-microgrid MILP-based model for DA scheduling. In Bahmani *et al.* (2020), MILP is used to formulate a cooperative market mechanism for energy transactions in a multi-microgrid setting. An RTP DR program is considered and each MG interconnects DGs, WTs, PVs, flexible and critical loads as well as ESS. Scenarios are generated in a stochastic way to account for market uncertainty and renewable energy generation. Various scenarios are investigated to include grid-connected operation, island mode and different connection levels between participating MGs. Results showed the effectiveness of this approach in lowering market prices and enhancing reliability in the case of increased power transaction capacity. Toward the application of state of the art, a microgrid energy management GA-based approach is applied in Wang *et al.* (2018) to optimize cost strategies for scheduling distributed energy resources. A quasi-static artificial bee colony approach is used to optimize a multiobjective DR problem based on the cost of energy and peak demand at the building level (Safamehr and Rahimi-Kian 2015), including PV, combined heat and power (CHP), batteries, electrical energy from the grid and natural gas. PSO is used in Carrasqueira *et al.* (2017) to solve bi-level problem modeling, the interaction between the retailer and consumers. The energy hub is explored in Batić *et al.* (2016) to develop a multicarrier DSM ToU optimization balancing energy import, conversion and storage. A multicarrier energy system of thermal, electrical and hydrogen loads is optimized using a fully decentralized multiagent approach (Shabani and Moghaddas-Tafreshi 2020). Comparing a case of responsive loads to a case without responsive loads via simulations led to the observation that DR could provide added value to the social welfare of the system and individual profit of agents.

Furthermore, a GA approach using present and DA data was tested by Ferrari *et al.* (2017) with respect to the management of loads of an experimental plant case study in Italy. The analysis involves PV, wind generation, a micro-CHP with a gas boiler and an absorption chiller coupled with thermal storage.

In addition, game theory is widely explored in formulating the interaction between consumers and utilities in DR. In Yang *et al.* (2018), this interaction is formed as a Bayesian game where the Bayesian Nash equilibrium is changed according to the regulation price set by the utility. Results indicate that this approach is effective in balancing energy and demand. Gong *et al.* (2019) developed a game-theoretic approach to test a distributed control strategy for large-scale DR consisting of high populations of EVs and storage devices. These flexible loads react to prices by optimizing their own objective functions in an agent-based framework. Prices are settled by solving a power flow dispatch optimization problem and results are presented to demonstrate the optimality of each individual's objective function and of social welfare for the system overall. According to this approach, full control is maintained at the customer side since it does not need to be transferred to the aggregator's side.

Abuelnasr *et al.* used a GA approach to evaluate the impact of different microgrid topologies on EMS operations considering energy losses, energy from the main grid and CO_2 emissions (Abuelnasr *et al.* 2018). Microgrid topologies consisted of networks, loads, biomass generation, PV and storages. More specifically, a GA optimization model is used to examine the influence of different microgrid topologies in energy management from the perspective of objective functions of energy loss, energy import and CO_2 emissions minimized individually. The participating microgrids are composed of 26 load points; 10 PV; DGs, half of which were of 200 kW and half of which were 500 kW; and a storage unit of 250 kW and 1000 kWh as well as biomass DGs. Eleven controls are employed in energy management optimization including four DR controls for loads above 50 kW, two three-phase dispatchable DGs, output power controls, three single-phase shunt capacitors control switches, one three-phase capacitor control switch and storage control. Three different configurations were investigated to identify optimal decisions and demonstrate the effectiveness of the proposed approach. A regional integrated energy system is modeled to optimize the operation of a system comprising wind power, concentrating solar power, gas power generation, thermal and electric power storage while meeting electric and thermal loads on the demand side (Jiang *et al.* 2020). The optimization model addresses (1) the total operational costs; (2) the system environmental cost; and (3) the system economic benefits due to participation in DR. The impact of different DR modes on system operation is considered using simulation. Results indicate that IDR can lead to a cost saving of 7.86% and a reduction of pollutants emissions of 18.37%. Furthermore, Alharbi and Bhattacharya (2018) present a stochastic EMS model to investigate several short-term dispatch probabilistic scenarios for isolated microgrids integrating wind and solar generation with EVs and DR.

1.3.6. *ANN-based short-term power forecasting*

ANN models are designed to imitate biological nervous system information processing and evolution. They have been used for years in different areas of engineering, science and business to deal with highly complex and nonlinear data sets. The ANN models assimilate the natural bonds of neurons and their high level interconnection to model complex systems. Artificial neural networks (ANN)-based short-term power forecasting is practiced to estimate DA loads and renewable energy production.

In the case of short-term predictions, the ANN models can be more effective compared to statistical, linear or nonlinear programming techniques. They encompass capabilities such as adaptive learning, self-organization, real-time operation, fault tolerance and the approximation of complex nonlinear functions. Kalaitzakis *et al.* (2002) tested advanced neural network short-term load forecasting using data from the electric power grid of the island of Crete in Greece. Various structures and configurations were assessed in their study and a parallel processing approach for a 24-h ahead prediction was demonstrated to be the most effective. ANN architectures for forecasting demand in electric power systems are presented by Tsekouras *et al.* (2015). A case study of the Greek electric power grid is used to explore the performance of different ANN configurations and factors, including period length and inputs for training, confidence interval and so on. Moreover, short-term power forecasting is of particular value for prosumers to model, understand and predict their consumption profiles, as well as to apply effective scheduling and control. A framework for district-level energy management and ANN forecasting at the building level was investigated by Hu *et al.* (2017), evaluating the performance for six buildings of different occupancy routines. Hybrid short-term load forecasting ANN combined with techniques such as fuzzy logic, GA and PSO are briefly discussed in Baliyan *et al.* (2015). Furthermore, a 24-h ahead prediction of excess power at the microgrid level is proposed by Mavrigiannaki *et al.* (2017), testing three different configurations, as a fundamental part of an advanced microgrid energy management strategy. Finally, an overview of load forecasting, dynamic pricing and DSM techniques in smart grid research applications reveals their potential for operational cost reductions between 5 and 25% (Khan *et al.* 2016).

1.4. Scientific focus of the book

Following the description of the state of the art, the scientific focus of the content presented in this volume is analyzed as follows:

– A comprehensive approach for evaluating the *performance gap of smart energy buildings / NZEBs* including industrial and residential case studies is developed. Dynamic energy models are developed, validated upon real data and exploited to explore key aspects of the operational behavior of buildings and systems. The developed approach provides an innovative, complete and transparent framework for analyzing the energy efficiency of buildings during their operational phase.

– *At the building level*, a novel *DR GA-based HVAC optimization scheme* is developed. According to this scheme, an optimization problem is formed to include the *cost of energy* and *predicted mean vote (PMV)* as the two criteria merged into one objective function. HVAC hourly set points are used as the variables of the optimization. A GA is used to identify dominant HVAC set point patterns based on dynamic energy prices, actual weather conditions and preferences with regard to indoor conditions. The developed approach constitutes a powerful assessment and decision tool, which can be used to identify and ultimately apply dominant HVAC set point patterns based on dynamic conditions. The GA optimization algorithm is coupled with the validated dynamic thermal model of the building enabling the assessment of energy cost, energy savings and thermal comfort for a wide range of temperature set point patterns and RTP schemes. The developed approach is explored to assess RTP schemes based on *real DA market information* on the basis of price fluctuations reflecting current market operations. This approach constitutes a significant contribution to the literature of HVAC energy management based on a DR perspective. According to the best of the author's knowledge, previous attempts to investigate this problem are limited to oversimplified mathematic models of the building HVAC operation. In addition, the innovation of the developed approach is justified by the fact that solutions are assessed against dynamic pricing profiles, which have been constructed based on real market data.

– At the district level, a DR energy management GA-based optimization approach based on day-ahead ANN-generated prediction models is proposed. The developed GA algorithm incorporates load shifting for the day ahead (24-h period) and evaluates possible alternatives based on cost and assumptions related to the practicality of the obtained solutions. The practical benefits of the proposed approach are linked to the development of a valuable tool for the evaluation of the potential rewards and risks of engagement in DR.

– The contribution of this work, at the district level, is related to linking ANN short-term electric forecasting and GA multiobjective optimization as a tool for generating and evaluating alternative DA load shifting solutions. In the case studies that follow, ToU and DA RTP schemes are assessed.

1.5. Book outline and objectives

The book structure and objectives are outlined below. In Chapter 2, the description of the facilities under investigation is provided. At the building level, these include Leaf House and Leaf Lab, a residential and an industrial smart energy building and ZEB, respectively. At the district level, these include the Leaf Community, a microgrid that includes several buildings (including the Leaf Lab), various renewable energy systems as well as thermal and electrical storage. In Chapter 3, a thorough analysis of the performance gap in one residential and one industrial smart NZEB is conducted. The analysis is based on a comparison of actual measured energy consumption during a full year period, with the energy performance according to the initial design considerations and a new developed dynamic simulation model. The model is developed based on the – as built – plans and a detailed building/systems audit. Subsequently, *in situ* measurements were obtained to record indoor temperature in representative thermal zones and use them in validating the building energy model. In Chapter 4, the validated building energy model of Leaf Lab was exploited in a novel DR HVAC GA optimization scheme. The optimization approach entailed the hourly temperature set point as a chromosome (decision variable) for the GA. The goal of the developed optimization model is to minimize daily HVAC energy cost while adhering to comfort standards. DA RTP profiles were created based on DA market information and results demonstrated significant margins of energy and cost savings throughout the year, especially when the daily variation of the pricing profiles allowed for adequate levels of load shifting between adjacent working hours. In Chapter 5, a new method for assessing DR energy management potential at district levels is presented. The method was developed and successfully tested to predict energy demand and optimize load shifting options to evaluate cost savings for the same energy consumption levels. An ANN-based algorithm is used for predicting DA consumption and a GA approach was implemented to provide balanced and cost optimum load shifting solutions.

Finally, the last part encompasses a critical reflection on key considerations and the main conclusions arising from this book. An overview of the limitations and constraints of the developed approaches is included along with future prospects recommendation for further work.

2

DR in Smart and Near-zero Energy Buildings: The Leaf Community

The Leaf Community is an industrial settlement owned and managed by the Loccioni Group for conducting research and innovation in the sectors of energy, the environment, automotives, aerospace, robotics and so on. The Leaf Community is a unique blend of inspired qualified personnel where the preservation of the natural environment, renewable energy sources (RES) and worldwide R&D meets education, local culture and society. Industrial buildings in the Leaf Community, located in Angeli di Rosora in the Italian province of Ancona, are mainly the key loads part of a microgrid interconnecting various photovoltaic (PV) installations, electric and thermal storage, micro-hydroelectric plants and electric vehicles (EVs). The climate in Ancona is Mediterranean, with hot, dry summers and mild winters. The warm season starts in June and lasts till mid-September

---

Chapter written by Nikos KAMPELIS, Konstantinos GOBAKIS, Vagias VAGIAS, Denia KOLOKOTSA, Laura STANDARDI, Daniela ISIDORI, Cristina CRISTALLI, Fabio Maria MONTAGNINO, Filippo PAREDES, Pietro MURATORE, Luca VENEZIA, Marina Kyprianou DRACOU, Alaric MONTENON, Andri PYRGOU, Theoni KARLESSI and Mattheos SANTAMOURIS. This chapter is based upon: "Evaluation of the performance gap in industrial, residential & tertiary near-zero energy buildings", *Energy and Buildings*, pp. 58–73, vol. 148, 2017.

with an average high temperature of 29°C and an average low temperature of 19°C. The cold season starts in November and ends in March with an average high temperature below 12°C (WeatherSpark n.d.).

The Leaf Community (Figure 2.1) consists of five industrial buildings (L3-AEA, L4-Leaf Lab, L5-Kite Lab and L6), one office building (L2-Summa) and a building used mainly for meetings (Leaf Farm). All buildings (except the Leaf Farm) are equipped with rooftop PVs of total power 629.2 kWp, and ground water heat pumps. In addition, a two-axis solar tracker of 18 kWp, a 48 kWp micro-hydro plant, a 224 kWh battery storage and a 523.25 kWh/K thermal storage are connected to the microgrid, which also features EV charging stations. Buildings, renewable energy systems (PVs, micro-hydro systems) and storage systems are all coupled and connected to the main power grid via a single interconnection line (the point of delivery).

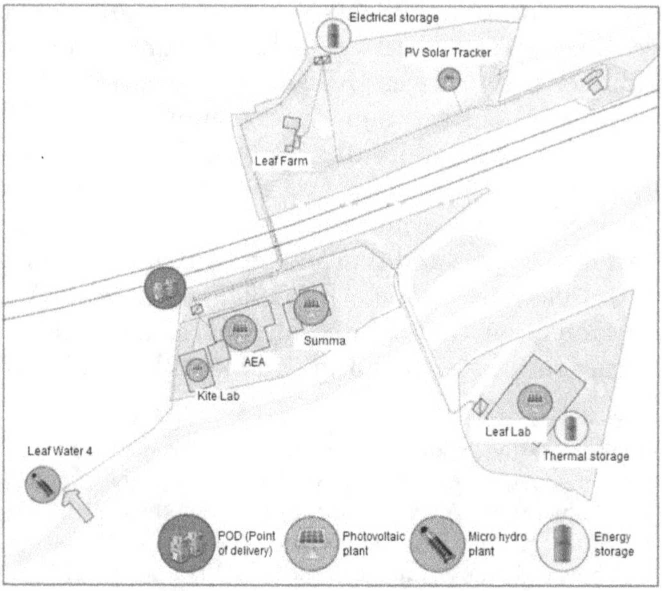

Figure 2.1. *The Leaf Community map. For a color version of this figure, see www.iste.co.uk/kolokotsa/smartbuildings.zip*

2.1. The Leaf Lab industrial building, AEA Italy

The Leaf Lab (Figure 2.2) is an industrial building of a rectangular shape and a floor area of approximately 6,000 m^2 located in the Leaf Community (Provata *et al.* 2015), one of the very well-established smart microgrids in Europe. The Leaf Lab incorporates the newest technology, making it exceptionally tolerant to external weather conditions. This reduces the amount of energy needed for heating, cooling, ventilation and lighting to a minimum. The Leaf Lab is a near-zero energy building (NZEB), combining passive systems, energy-efficient technologies, integrated monitoring and control, as well as renewable energy production. Renewable energy is exploited with the use of PV systems, thermal storage and heat pumps. Thermal storage is exploited to optimize heating, ventilation and air conditioning (HVAC) performance and minimize dependency on energy imported from the main grid.

Figure 2.2. *The Leaf Lab*

The building envelope of the Leaf Lab consists of highly insulated external walls with U-value of 0.226 W/m^{2}K and double-glazed windows with U-value between 1.793 and 3.194 W/m^{2}K. The HVAC system installed in the Leaf Lab is composed of heat pumps with a heating coefficient of performance (COP) of 4.8 and cooling energy efficiency ratio (EER) of 6.2–7. A thermal storage water tank of 400 m^3 is coupled to the building's HVAC system to reduce peak power and to improve the efficiency of the HVAC system during heating and cooling periods throughout the year (Kolokotsa *et al.* 2019). This is implemented using excess energy from the PV, i.e. during weekends and holidays, to operate the heat pumps and store heating or cooling energy in the thermal tank. Stored energy is then used to optimize the HVAC efficiency by reducing peak demand and imported energy consumption during working hours. The HVAC is controlled with digital thermostats distributed in the various thermal zones, satisfying the set point limits of the CEN 15251 (20–24°C for heating, 23–26°C for cooling). The set points for the industrial and office spaces in heating mode are 21°C and 22°C, whereas in cooling mode the set points are 25°C and 26°C, respectively.

Illuminance sensors, controlling artificial lighting in the indoor spaces of the Leaf Lab, activate dimmable LED lights when levels of natural lighting fall below 500 lux. Furthermore, automated shading is installed in the vast majority of the windows and operated according to the altitude of the sun. This allows for natural light to be adequately leveled for visual comfort throughout clear sky days while minimizing energy consumption for artificial lighting and avoiding glare. Finally, as shown in Figure 2.2, a rooftop PV system of 236.5 kWp is installed in the Leaf Lab.

The energy efficiency of the Leaf Lab as recorded in the energy certificate was A+, associated with net primary energy consumption of 4.11 kWh/m^3 (equivalent to 26.91 kWh/m^2).

2.2. The Leaf House residential building, AEA Italy

The Leaf House (Figure 2.3) is a residential building of exceptional bioclimatic design and smart technologies (Comodi *et al.* 2015). It consists of six highly insulated apartments with a total floor area of approximately 470 m^2, a ventilated roof, solar tubes, smart monitoring and controls, building-integrated PVs, geothermal air preconditioning with heat pumps, solar thermal collectors, electrical storage and a user-friendly energy management system for residents. The number of residents in the Leaf House varies, as it accommodates both employees of the Loccioni Group (2017) and short-term visitors of the Leaf Community (n.d.).

Figure 2.3. *The Leaf House*

The building envelope of the Leaf House is composed of external walls with a U-value of 0.41 W/m^2K and windows of total U-value between 0.73 and 1.49 W/m^2K. The HVAC system installed in the Leaf House is composed of three heat pumps with geothermal air preconditioning and heat recovery, connected to a radiant floor distribution system. The heating COP of the heat pumps is in the range of 2.9 to 4.6, while the cooling EER varies between 1.9 and 3.6. Seven solar thermal collectors, of a total area of 19 m^2, are connected to a 1 m^3 thermal storage boiler of 15 kW electrical power for domestic hot water and space heating. Moreover, 115 PV panels and a total peak power of 20 kWp covering a total area of 150 m^2 are integrated into the Leaf House's rooftop, as depicted in Figure 2.3. The energy produced by the PV system is mainly exploited to power the geothermal heat pumps and reduce the overall power consumption. According to the energy certificate of the Leaf House, its normalized annual primary energy consumption is 19.66 kWh/m^2, corresponding to an energy efficiency class of A+. The apartments in the Leaf House are equipped with a touch display providing access to an energy management interface for observing indoor conditions and energy-related data as well as managing the HVAC, lights, window shutters, etc. An extensive database of measurements for each apartment in the Leaf House, including power-related data, is also accessible online, restricted to authorized use only through the MyLeaf platform.

3

Performance of Industrial and Residential Near-zero Energy Buildings

In this chapter, a comprehensive approach for evaluating the performance of one industrial and one residential smart/near-zero energy building (NZEB) is presented. Initially, the buildings are audited for a detailed investigation of their construction characteristics, installed systems and controls. Subsequently, holistic data from advanced metering and sensor equipment is explored to verify energy demand and actual performance. Dynamic energy models are developed, validated and tested to explore key aspects of the operational behavior of buildings and systems and to draw out essential knowledge about their performance. A comparison of measured data with dynamic modeling results and the initial design energy efficiency certification study is explored to

---

Chapter written by Nikos KAMPELIS, Konstantinos GOBAKIS, Vagias VAGIAS, Denia KOLOKOTSA, Laura STANDARDI, Daniela ISIDORI, Cristina CRISTALLI, Fabio Maria MONTAGNINO, Filippo PAREDES, Pietro MURATORE, Luca VENEZIA, Marina Kyprianou DRACOU, Alaric MONTENON, Andri PYRGOU, Theoni KARLESSI and Mattheos SANTAMOURIS. This chapter is based upon: “Evaluation of the performance gap in industrial, residential & tertiary near-zero energy buildings”, *Energy and Buildings*, pp. 58–73, vol. 148, 2017.

address the actual performance gap, reflect on the limitations of each approach and highlight important conclusions arising from this work.

3.1. Materials and methods

The research activities performed and presented in this chapter target the estimation and evaluation of the performance gap between the design and operational phase of zero energy buildings.

The steps are as follows:

1) Selection of the case study buildings: two case studies are analyzed to cover industrial and residential building use. The Leaf Lab and the Leaf House envisage unique building designs for minimizing net energy consumption. This is achieved through a variety of measures, including responsive building envelope applications, efficient HVAC systems coupled with thermal storage, intelligent controls, renewable energy systems and integrated energy management. The initial design target of the two buildings, to operate at near-zero energy, is established based on their energy certification process and is used throughout the text as a working definition serving qualitative assessment purposes.

2) The second step involves an analysis of the buildings and their systems' design and an assessment of power and energy requirements through dynamic thermal simulation models.

3) The third step is the data collection while the buildings are in operation. To test and evaluate:

i) the performance gap between the developed dynamic simulation models and actual operation;

ii) the performance gap between the initial zero energy targets and buildings' actual operation.

4) The fourth step includes a comparison of the results of the buildings and the extraction of useful remarks and conclusions.

In our analysis, a combination of metrics, including primary energy consumption and end-use net consumption (absolute and normalized), as well as $CO_{2\text{-eq}}$ emissions, is deployed. The period of balance is annual, to account for yearly representative thermal loads and renewable energy production. Renewable supply in the considered cases is on-site and building-integrated. Of the examined cases, the residential building is directly connected to the power distribution grid and the industrial building as part of the Leaf Community microgrid.

3.1.1. *Energy simulation model*

EnergyPlus is the simulation engine software used to conduct an integrated simulation of the building, system and plant, whereby supply and demand are matched, based on successive iteration substitution following Gauss–Seidel updating (U.S. Department of Energy 2019). OpenStudio is used as the API software for developing and parameterizing the model, following the principles outlined by Brackney *et al.* (2018). Ambient temperature, relative humidity, solar radiation and wind speed data were obtained from local meteorological equipment and converted to a yearly weather file. The data of total HVAC energy consumption is exploited to provide the baseline against which the model-based results are evaluated.

The simulation model contains, on the one hand, the geometry, construction components and materials of the building under study. For opaque material thickness (m), the thermal conductivity (W/mK), density (kg/m^3) and thermal absorptance (dimensionless) properties are edited. For transparent materials, such as glass in windows and sky windows thickness (m), the thermal conductivity (W/mK) and optical properties, such as solar, visible and infrared transmittance, are inserted. On the other hand, a model of the HVAC system is designed based on the installed technologies and adjusted according to the actual key performance heat pump technical parameters, such as coefficients of performance (COP), fan maximum flow power (m^3/s), pressure rise and efficiency. Other parameters, such as rated total heating/cooling capacity and rated and maximum air flow rated are automatically sized, based on the software's calculations. Furthermore, with respect to the operation of the major installed systems, the simulation model takes into account the temperature set points of the HVAC system, the ventilation and infiltration rates (ACH^{-1}) and a number of schedules to determine the artificial lighting, electric equipment and occupancy. Subsequently, an intensive search of the parameters that affected the daily, monthly and annual power distribution profiles is carried out to improve the initial results of the model by minimizing the deviation from HVAC power consumption data. Through trial and error, various combinations and fine-tuning of all the above parameters are carried out to reach the optimum results when assessing intraday, monthly and annual deviation levels.

EnergyPlus simulation is based on heat balance calculations solved simultaneously with the aid of an integration solution manager, which includes surface heat balance, air heat balance and building systems simulation blocks. The heat balance of outside surfaces is calculated based on the following equation:

$$q''_{asol} + q''_{LWR} + q''_{conv} - q''_{ko} = 0 \qquad [3.1]$$

where q''_{asol} is the absorbed direct and diffuse solar (short wavelength) radiation and heat flux, q''_{LWR} is the net long wavelength (thermal) radiation flux exchange with the air and surroundings, q''_{conv} is the convective flux exchange with the outside air and q''_{ko} is the conduction heat flux (q/A) into the walls.

Clearly, q''_{asol} is influenced by parameters such as location, surface angle and tilt, surface material and weather conditions. q''_{LWR} is determined by the radiation exchange between the surface and the ground, sky and air. It is dependent on the absorptivity and emissivity of the surface; the temperature of the surface, sky, ground and air and corresponding view factors. Assumptions, such as each surface is at a uniform temperature and the energy flux leaving a surface is evenly distributed, are considered reasonable for building an energy simulation. Using the Stefan–Boltzmann law in the above equation yields:

$$q''_{LWR} = \epsilon\sigma F_{gnd}\left(T_{gnd}^4 - T_{surf}^4\right) + \epsilon\sigma F_{sky}\left(T_{sky}^4 - T_{surf}^4\right) + \epsilon\sigma F_{air}\left(T_{air}^4 - T_{surf}^4\right) \quad [3.2]$$

where ϵ is the long-wave emittance of the surface, σ is the Stefan–Boltzmann constant, F_{gnd} is the view factor of wall surface to ground surface temperature, F_{sky} is the view factor of wall surface to sky temperature, F_{air} is the view factor of wall surface to air temperature, T_{surf} is the outside surface temperature, T_{gnd} is the ground surface temperature, T_{sky} is the sky temperature and T_{air} is the air temperature.

The above equation is converted by introducing linear radiative heat transfer coefficients such that:

$$q''_{LWR} = h_{r,gnd}\left(T_{gnd} - T_{surf}\right) + h_{r,sky}\left(T_{sky} - T_{surf}\right) + h_{r,air}\left(T_{air} - T_{surf}\right) \quad [3.3]$$

where

$$h_{r,gnd} = \epsilon\sigma F_{gnd}\left(T_{surf}^4 - T_{gnd}^4\right)/(T_{surf} - T_{gnd})$$

$$h_{r,sky} = \epsilon\sigma F_{sky}\left(T_{surf}^4 - T_{sky}^4\right)/(T_{surf} - T_{sky})$$

$$h_{r,air} = \epsilon\sigma F_{air}\left(T_{surf}^4 - T_{air}^4\right)/(T_{surf} - T_{air})$$

Exterior convection is modeled using the following equation:

$$q''_{conv} = h_{c,ext}A(T_{surf} - T_{air}) \quad [3.4]$$

where q''_{conv} is the rate of exterior convective heat transfer, $h_{c,ext}$ is the exterior convection coefficient, A is the surface area, T_{surf} is the surface temperature and T_{air} is the outdoor air temperature.

Conduction heat fluxes are modeled based on the following equation:

$$q''_{ko}(t) = \sum_{j=0}^{\infty} X_j T_{0,t-j\delta} - \sum_{j=0}^{\infty} Y_j T_{i,t-j\delta} \quad [3.5]$$

where $q''_{ko}(t)$ is the conductive heat flux for the current time step, T is temperature, i indicates the internal element of the building, o indicates the external element of the building and X, Y are the response factors.

In more detail, conduction transfer functions (CTFs) as shown below are used to estimate the heat fluxes on either side of the building elements, based on previous temperature values of interior and exterior surfaces, as well as previous interior flux values.

$$q''_{ki}(t) = -Z_o T_{i,t} - \sum_{j=1}^{nz} Z_j T_{i,t-j\delta} + Y_o T_{o,t} + \sum_{j=1}^{nz} Y_j T_{o,t-j\delta} + \sum_{j=1}^{nz} \Phi_j q''_{ki,t-j\delta} \quad [3.6]$$

$$q''_{ko}(t) = -Y_o T_{i,t} - \sum_{j=1}^{nz} Y_j T_{i,t-j\delta} + X_o T_{o,t} + \sum_{j=1}^{nz} X_j T_{o,t-j\delta} + \sum_{j=1}^{nz} \Phi_j q''_{ko,t-j\delta} \quad [3.7]$$

where X_j is the outside CTF coefficient, j = 0, 1,...,nz, Y_j is the cross CTF coefficient, j = 0, 1,...,nz, Zj is the inside CTF coefficient, j = 0, 1,...,nz, Φ_j is the flux CTF coefficient, j = 0, 1,...,nq, T_i is the inside surface temperature, T_o is the outside surface temperature, q''_{k0} is the conduction heat flux on the outside face and q''_{ki} is the conduction heat flux on the inside face.

In addition, for each thermal zone, EnergyPlus simulation is based on an integration of energy and moisture balance as shown in the following equation:

$$C_z \frac{dT_z}{dt} = \sum_{i=1}^{N_{sl}} \dot{Q}_i + \sum_{i=1}^{N_{surfaces}} h_i A_i (T_{si} - T_z) + \sum_{i=1}^{N_{surfaces}} m_i C_p (T_{zi} - T_z) + m_{inf} C_p (T_\infty - T_z) + \dot{Q}_{sys} \quad [3.8]$$

where $\sum_{i=1}^{N_{sl}} \dot{Q}_i$ is the sum of the convective heat transfer from the zone surfaces, $\sum_{i=1}^{N_{surfaces}} h_i A_i (T_{si} - T_z)$ is the convective heat transfer from the zone surfaces, $m_{inf} C_p (T_\infty - T_z)$ is the heat transfer due to the infiltration of outside air, $\sum_{i=1}^{N_{surfaces}} m_i C_p (T_{zi} - T_z)$ is the heat transfer due to interzone air mixing, $\dot{Q}_{sys}$ is the air systems output, $C_z \frac{dT_z}{dt}$ is the energy stored in the zone air and T_∞ is the fluid temperature (K).

And

$$C_z = \rho_{air} C_p C_T \quad [3.9]$$

where ρ_{air} is the zone air density, C_p the zone air specific heat and C_T the sensible heat capacity multiplier.

Infiltration is defined as the outdoor air unintentionally entering the building due to doors being opened as well as air leakage through windows and other openings. Infiltrated air is mixed with air in the various thermal zones of the building. Determining infiltration (or air tightness) values contains significant uncertainty, as it requires a complex and elaborate procedure, often referred to as blower door test. Infiltrated air is commonly modeled as the number of air changes per hour (ACH^{-1}) and taken into account in the air heat balance at a temperature equal to that of ambient air. In EnergyPlus, infiltration is modeled based on the following equation:

$$Infiltration = (I_{design})(F_{schedule})[A + B|(T_{zone} - T_{odb})| + C(Windspeed) + D(Windspeed^2)] \quad [3.10]$$

where I_{design} is the user-defined infiltration value (ACH^{-1}), T_{zone} is the zone air temperature at current conditions (°C), T_{odb} is the outdoor air dry-bulb temperature (°C), $F_{schedule}$ is the user-defined schedule value between 0 and 1, A is the constant term coefficient, B is the temperature term coefficient, C is the velocity term coefficient and D is the velocity squared coefficient.

Similarly, ventilation can be modeled using a schedule, maximum and minimum values and delta temperature values, and is determined by the following equation:

$$Ventilation = (V_{design})(F_{schedule})[A + B|(T_{zone} - T_{odb})| + C(Windspeed) + D(Windspeed^2)] \quad [3.11]$$

where V_{design} is the user-defined ventilation value (ACH^{-1}), T_{zone} is the zone air temperature at current conditions (°C), T_{odb} is the outdoor air dry-bulb temperature (°C), $F_{schedule}$ is the user-defined schedule value between 0 and 1, A is the constant term coefficient, B is the temperature term coefficient, C is the velocity term coefficient and D is the velocity squared coefficient.

Furthermore, the energy provided to each thermal zone by the HVAC system $\dot{Q}_{sys}$ is given by:

$$\dot{Q}_{sys} = \dot{m}_{sys} C_p (T_{sys} - T_z) \quad [3.12]$$

Equations [3.8] and [3.12] can be transformed to yield zone air temperature, as shown in the following equation:

$$T_z^t = \frac{\sum_{i=1}^{N_{sl}} \dot{Q}_i^t + \dot{m}_{sys} C_p T_{supply}^t + \left(C_z \frac{T_z}{\delta t} + \sum_{i=1}^{N_{surfaces}} h_i A_i T_{si} + \sum_{i=1}^{N_{zones}} \dot{m}_i C_p T_{zi} + \dot{m}_{inf} C_p T_\infty \right)^{t-\delta t}}{\frac{C_z}{\delta t} + \left(\sum_{i=1}^{N_{surfaces}} h_i A_i + \sum_{i=1}^{N_{zones}} \dot{m}_i C_p + \dot{m}_{inf} C_p + \dot{m}_{sys} C_p \right)} \quad [3.13]$$

3.2. Energy performance analysis

3.2.1. *The Leaf Lab*

The aim of this section is to analyze the Leaf Lab's energy performance and compare modeling results with real-time data. Modeling and simulation for the Leaf Lab are carried out using Google SketchUp (2017) as the graphical user interface for 3D modeling, OpenStudio (2017) plugin and standalone application for editing the various model parameters and EnergyPlus (U.S. Department of Energy 2015) as the simulation engine. The model is depicted in Figure 3.1. Architectural drawings are used to design the building structure and envelope as well as to convert the spaces into thermal zones. Electromechanical and implemented HVAC system designs are taken into consideration. Moreover, the physical and thermal characteristics of the external and internal walls, roof, ground floor and ceiling, alongside similar information about the external windows, are collected. The lights of each space, approximate number of people in each space and equipment information are recorded for the estimation of the internal thermal gains and electrical energy consumption.

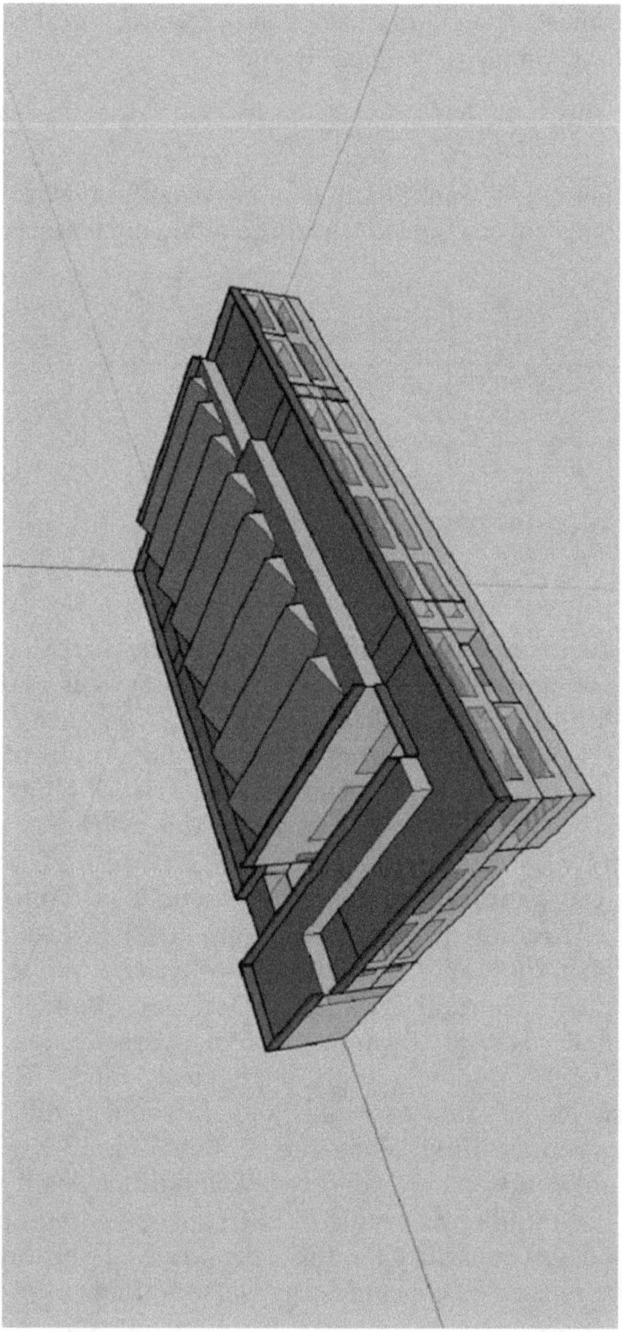

Figure 3.1. *The model of the Leaf Lab in Google SketchUp. For a color version of this figure, see www.iste.co.uk/kolokotsa/smartbuildings.zip*

Energy consumption and production data from measurements is collected, analyzed and processed to serve the scope of the analysis. The validation of the model is then performed using data-recorded temperature and relative humidity sensors, installed carefully in selected representative thermal zones of the building, taking into consideration size, orientation, use and contact with the ground or outdoor air. Additionally, data is extracted by MyLeaf (2018), a specialized Loccioni Group proprietary web-based energy management system (EMS), providing reliable and user-friendly representation of any energy-related monitored parameters, such as ambient and indoor environment conditions, power consumption, production and storage over time. The open MyLeaf architecture allows the integration of advanced energy management and control applications at both the building and microgrid (district) level.

Specifically, the collected data from MyLeaf is (1) the building total and HVAC power demand; and (2) the power production by the photovoltaic (PV) system.

As can be observed in Figures 3.2 and 3.3, the simulated indoor temperature versus the measured one is less than 1 K at all times. This applies to the reception area as well as to all the rooms monitored, indicating high levels of agreement between the simulated and measured data.

A comparison of the measured and simulated energy consumption is shown in Table 3.1. We can observe that the difference in energy consumption between the various categories is 1.4% for artificial lighting, 0.6% for HVAC, 0.4% for equipment (including industrial processes) and 0.1% in total, demonstrating a strong correlation between the simulation results and the actual behavior of the building during its operational phase.

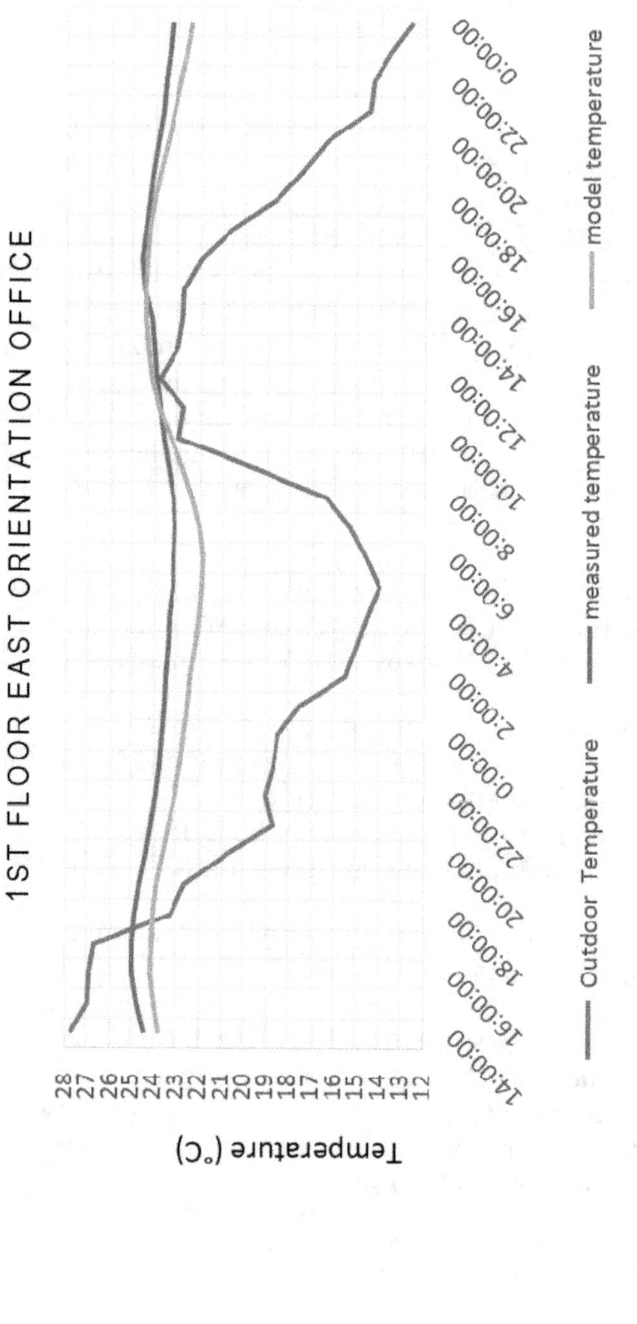

Figure 3.2. *First floor, east office, measured and simulated indoor temperature. For a color version of this figure, see www.iste.co.uk/kolokotsa/smartbuildings.zip*

Figure 3.3. *Ground floor, Leaf Lab reception, measured and simulated indoor temperature. For a color version of this figure, see www.iste.co.uk/kolokotsa/smartbuildings.zip*

Leaf Lab (industrial)		Artificial lighting	HVAC	Industrial/office equipment	Total
Monitored data	Electrical energy consumption (kWh)	35,467.3	227,176.1	297,366.1	560,009.5
	Normalized electrical energy consumption (kWh/m^2)	5.9	37.9	49.6	93.3
	Energy consumption (%)	6.3%	40.6%	53.1%	100.00%
	Normalized primary energy consumption (kWh/m^2)	11.0	70.4	92.2	173.6
	Energy production by the PV (kWh)				275,942
	Normalized PV energy (kWh/m^2)				46
Simulated data	Energy consumption (kWh)	34,985.5	225,838.3	298,604.2	559,428.0
	Normalized electrical energy consumption (kWh/m^2)	5.8	37.6	49.8	93.2
	Energy consumption (%)	6.3%	40.4%	53.4%	100.0%
Difference	Energy consumption (kWh)	481.8	1,337.8	–1,238.1	581.5
	Energy consumption (%)	1.4%	0.6%	0.4%	0.1%

Table 3.1. *Validation of the Leaf Lab model based on data from MyLeaf*

As indicated in Table 3.1, the energy consumption share of the industrial/office operations in the Leaf Lab is the highest of the categories, accounting for 53.1%. This is of particular importance when one considers the energy balance (especially given the PV electrical energy production of 46 kWh/m^2) as it reveals the HVAC and lighting systems, electrical energy consumption being equal to 43.8 kWh/m^2. For the conversion of electrical energy to primary energy consumption, a factor of 1.86 is used, based on internationally reported calculations for the energy mix and power grid efficiency of Italy (EEAP 2014). Taking into account energy production from the PV plant, it is concluded that the Leaf Lab is a near-zero energy industrial building with total net electrical energy consumption of 47.3 kWh/m^2 and normalized total net primary energy consumption of 127.6 kWh/m^2.

The correlation of the Leaf Lab model and the measured HVAC power demand on a monthly basis, as presented in Figure 3.4, demonstrates part of the validation process according to standardized measurement and verification principles (Webster *et al.* 2015). In the examined case, the coefficient of variation (C_v) of the root mean square error (RMSE) of 14.8% satisfies the International Performance Measurement and Verification Protocol (IPMVP) acceptable monthly tolerance levels.

3.2.2. *The Leaf House*

Modeling and analysis of the Leaf House, as in the case of the Leaf Lab, is carried out using Google SketchUp (2017) as the graphical user interface for 3D modeling, OpenStudio (2017) plugin and standalone application for editing the various model parameters and EnergyPlus (U.S. Department of Energy 2015) as the simulation engine. The developed 3D model is depicted in Figure 3.5. The thermal zone division is performed with large attention to detail to best capture differences in indoor comfort, leading to every room being considered a separate thermal zone.

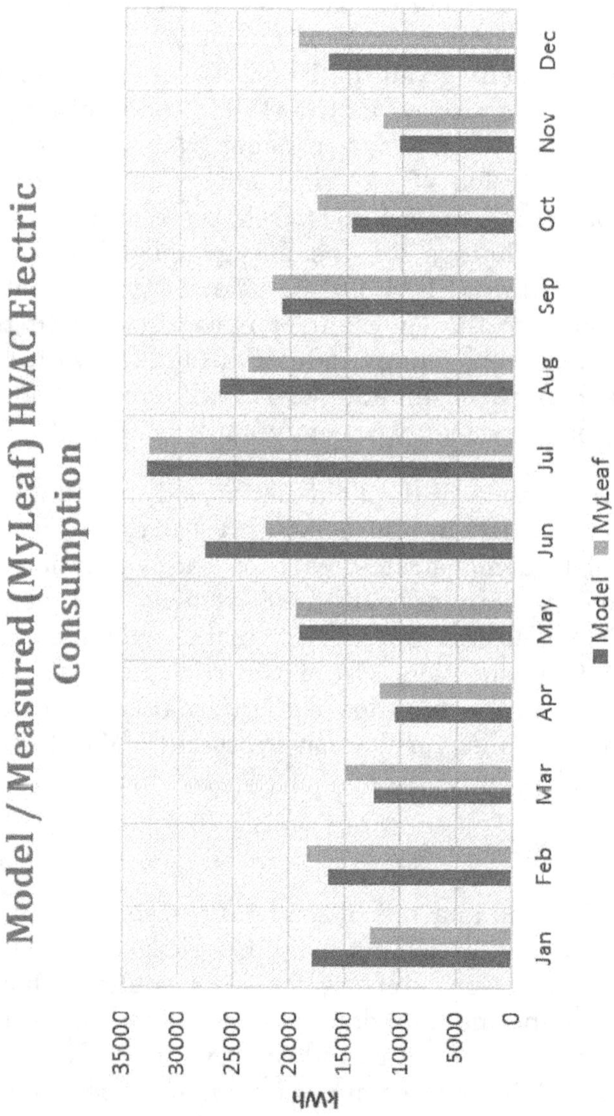

Figure 3.4. *HVAC system validation based on monthly electrical energy consumption. For a color version of this figure, see www.iste.co.uk/kolokotsa/smartbuildings.zip*

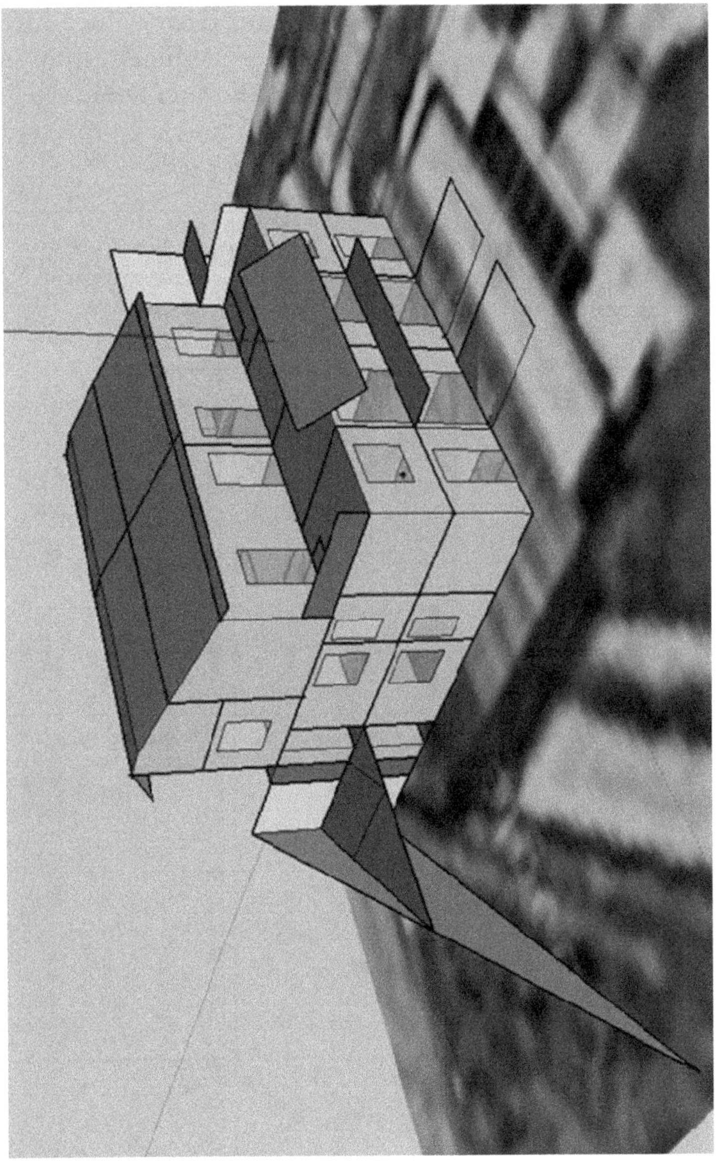

Figure 3.5. *The Leaf House and its thermal energy model using OpenStudio plugin. For a color version of this figure, see www.iste.co.uk/kolokotsa/smartbuildings.zip*

Energy performance in the Leaf House, according to 2015 data from MyLeaf, is summarized in Table 3.2. In the measurements, it is observed that the Leaf House is a NZEB, because its normalized primary energy consumption is 54.4 kWh/m^2. The PV system energy production accounts for 63.1% of the building energy demand and a $CO_{2\text{-eq}}$ emissions reduction of 11.32 t on a yearly basis (Figure 3.6).

Leaf House	Total	Total net (consumption minus production)
Annual electrical energy consumption (kWh)	37,196.0	13,746.0
Normalized annual electrical energy consumption (kWh/m^2)	79.1	29.2
Primary annual energy consumption (kWh)	69,184.6	25,567.6
Normalized annual primary energy consumption (kWh/m^2)	147.2	54.4
Annual $CO_{2\text{-eq}}$ emissions (kg)	17,965.7	6,639.3

Table 3.2. *Leaf House energy consumption data for 2015 (MyLeaf)*

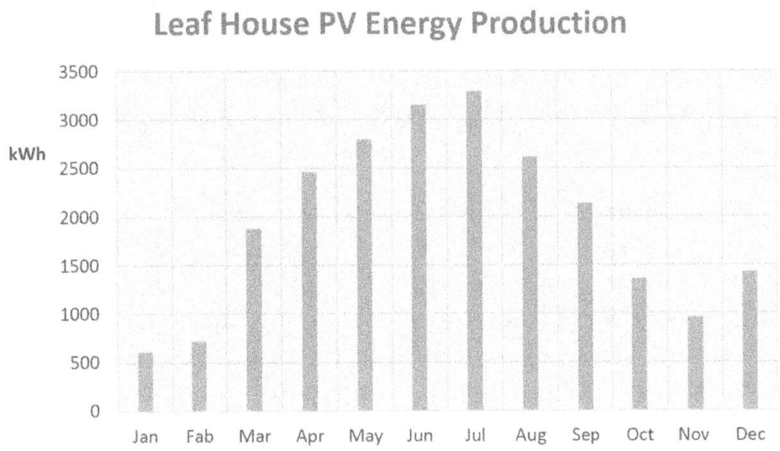

Figure 3.6. *Leaf House PV system monthly energy production for 2015 (MyLeaf)*

3.3. Discussion

In the selected case studies, the performance gap is finally assessed by comparing design and operational primary energy consumption, as presented in Table 3.3.

Pilot case study	Normalized primary energy consumption in design phase (kWh/m^2)	Normalized net primary energy consumption in operational phase (kWh/m^2)
Leaf Lab	26.9	35.4
Leaf House	19.6	54.4

Table 3.3. *Normalized primary energy consumption in the design and operational phase*

With regard to the Leaf Lab, the net normalized primary energy in the operational phase is calculated by deducing the energy dissipated for industrial purposes, as this is not taken into consideration in the corresponding design value. According to the results, there is a relatively low difference of 8.5 kWh/m^2 in primary energy consumption, which is not considered particularly significant.

In the case of the Leaf House, the performance gap is of higher magnitude, to be specific, 34.8 kWh/m^2 of primary energy consumption. A possible explanation for – at least – part of this performance gap is that the energy classification process in Italy (as well as in other countries) does not take into account energy for lighting or other appliances, as it depends on residents' behavior or other factors that cannot be standardized and applied as a common assessment framework. One issue that may also be related to the performance gap in this case is associated with the operation of the hydronic underfloor heating in the Leaf House. The system is characterized by high thermal inertia, which is slow in responding to weather changes. In this regard, it would be interesting to evaluate alternative advanced controls (i.e. predictive control) effectiveness in improving energy efficiency and indoor comfort levels. Another critical consideration with respect to the performance gap in the Leaf House concerns the engagement of residents, in terms of their ability in controlling building systems, their understanding of the actual potential in saving energy and their motivation in this direction. Despite the fact that residents of the Leaf Lab enjoy an elaborate monitoring and control interface, it has not been adequately explored if a performance gap may be linked to a lack of capability in using the elegant controls provided or a low commitment to addressing energy savings. An important parameter in this direction is that residents in the Leaf Lab are often visitors who do not permanently reside in the building but in an *ad hoc* fashion.

Overall, in the examined cases, the performance gap is either not particularly significant or can be possibly addressed by technical improvements or factors related to human activity. In the case of the Leaf Lab and the Leaf House, this is largely due to the integrated initial design, involving implementation of state-of-the-art techniques, technologies and know-how for achieving near-zero energy goals. In the case of the Leaf House, technical measures, such as predictive control, could possibly provide a smart solution to avoiding energy waste and improving indoor conditions. On the other hand, training about the available controls, behavioral change and active engagement can be especially important for residents to become proactive in reducing energy consumption to even lower levels. Behavioral change can be achieved in a number of ways, including raising awareness and gamification, i.e. competitions between apartments or enrolment in rewarding (future) demand response programs.

3.4. Conclusion

In this chapter, the operational performance of industrial and residential buildings has been investigated, analyzed and optimized with the use of dynamic simulation tools. Energy-efficient technologies, renewable energy technologies, storage and smart monitoring and controls have been audited to evaluate their significance for smart NZEBs of different types. Various performance indicators have been used in this analysis, including normalized electrical and primary energy consumption. Smart monitoring and indoor conditions' measurements have been used to allow the extraction of robust results and the validation of dynamic building energy models. The above analysis reveals the significance of evaluating the actual performance gap in NZEBs and provides the basis for decision-making and smart adjustments as necessary. In both cases, apart from the high-quality building envelopes, the near-zero target is largely pursued by renewable energy technologies and the implementation of advanced monitoring and controls. Furthermore, in the aforementioned cases, there is a systematic and continuous approach in establishing near-zero energy targets, through research and innovation activities. In this direction, predictive control, behavioral change and proactive users' engagement through gamification and enrolment in demand response programs have been identified as potential areas for addressing energy efficiency improvements in the future.

4

HVAC Optimization Genetic Algorithm for Industrial Near-Zero Energy Building Demand Response

Demand response (DR) makes it possible to alter the profile of power consumption in individual buildings and building districts, i.e. microgrids, for economic return. There is significant potential for DR in enabling flexibility via advanced grid management options, allowing higher renewable energy penetration and the efficient exploitation of resources. DR and the dynamic management of distributed energy resources are gradually gaining importance as valuable assets for managing peak loads, grid balance, renewable energy source intermittency and energy losses. In this chapter, the potential for operational optimization of a heating, ventilation and air conditioning (HVAC) system in a smart near-zero energy industrial building is investigated with the aid of a genetic algorithm (GA). The analysis involves the validated building energy model of the Leaf Lab presented in Chapter 3, a model of energy costs and an optimization model for establishing HVAC optimum temperature set points. Optimization aims to

---

Chapter written by Nikos KAMPELIS, Nikolaos SIFAKIS, Denia KOLOKOTSA, Konstantinos GOBAKIS, Konstantinos KALAITZAKIS, Daniela ISIDORI and Cristina CRISTALLI. This chapter is based on: "HVAC optimisation algorithm for industrial near-zero energy building demand response", *Energies*, 2019.

establish a trade-off between the minimum daily cost of energy and thermal comfort. The predicted mean vote (PMV) is integrated into the objective function to ensure that thermal comfort requirements are met.

The purpose of this chapter is to propose a GA optimization approach and to investigate its effectiveness in HVAC temperature set point control, based on day-ahead pricing information, and using profits as a reward for exploiting flexibility. The cost of energy is used as one of the two optimization criteria and is naturally, as well as in this case, a function of energy consumption over time. Using energy consumption as the optimization criterion instead, would lead to suboptimum performance with respect to cost, which is the main incentive behind changes in power consumption. Most importantly, minimizing on-site energy consumption measured at the point of consumption does not guarantee optimum environmental performance, since it does not take into account where, when and how energy is generated. On the other hand, having the cost of energy as one criterion in the objective function provides an indirect way to account for operational aspects of the power grid, provided that the energy market allows the penetration of DR resources, as well as distributed renewable energy generation. The reduction of on-site energy consumption is in this case, considered as an indirect goal and is evaluated, since it is acknowledged as a well-established measure providing necessary information on the energy efficiency of buildings, and cannot be neglected.

4.1. Methodology

The framework presented hereafter concerns the optimization of the HVAC temperature set point hourly schedule based on a GA, incorporating daily operational cost and the mean PMV as the two criteria of the objective function. Operational cost refers to the cost of energy on the basis of the given day-ahead hourly pricing profile and the HVAC hourly energy consumption, which is obtained by the simulation of the building's validated model. The building thermal model is validated based on annual HVAC energy consumption and the measurement of indoor temperature (Kampelis *et al.* 2017). The validated thermal model of the building provides a reliable basis for this kind of investigation, as it takes into consideration the physical aspects of the building (geometry, materials), operational aspects and climate conditions in a dynamic way. The baseline scenario is a reliable benchmark which the optimized scenario is compared against. Therefore, operational effects are kept constant to account for the fact that user behavior, natural ventilation and industrial operations are difficult to model and in most cases, are not monitored. On the other hand, inducing changes in the temperature set points of the HVAC system makes it imperative to evaluate any solution on the basis of the building users' thermal sensation, and the heat exchange between the human body and the surrounding indoor environment. This balance of energy fluxes is influenced by physical activity, clothing and the following indoor conditions: air temperature, mean radiant temperature, air velocity and relative humidity (RH). Internal comfort is evaluated in this work using the PMV, index as developed by Fanger in 1972 and adopted by the International Organization for Standardization in Standard ISO 7730:2005, to account for human heat generation and exchange with the surrounding environment (International Organization for Standardization 2006). PMV is converted to the percentage of people dissatisfied (PPD) to provide an estimate of the amount of people who feel uneasy with certain thermal conditions. The decision variables in the optimization process are the hourly HVAC temperature set points. Controlling the HVAC temperature set points has, as a consequence, caused variations in the operation, power consumption and running cost of the HVAC system. Naturally, this will impact indoor thermal conditions, thereby imposing the need for including thermal comfort as a criterion into the optimization process and compliance with established standards.

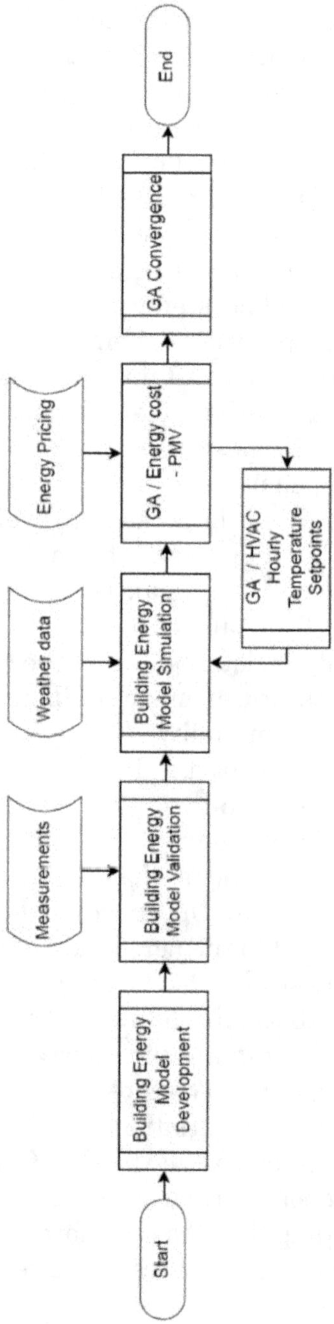

Figure 4.1. *Genetic algorithm (GA)-based heating, ventilation and air conditioning (HVAC) temperature set point optimization scheme*

The methodology followed is depicted in Figure 4.1. Firstly, the three-dimensional (3D) thermal model of the building (Leaf Lab) was developed in OpenStudio, based on the technical information of the building (i.e. drawings and datasheets of systems installed) and site audits. Secondly, the model was validated using measurements of weather conditions, indoor (air temperature, RH) conditions and HVAC power consumption. Details on the building modeling and validation procedures are available in Kampelis *et al.* (2017). Thirdly, a new weather file was constructed for the year of interest by merging together weather measurements, including dry and wet bulb temperature (°C), atmospheric pressure (kPa), RH (%), dew point temperature (°C), global, normal and diffuse solar irradiance (Wh/m^2) and wind speed (m/s). The validated 3D thermal model of the building was set up to receive an input of the temperature set points from an external source (Matlab in this case) when simulating the building's energy performance, and provide HVAC power demand (P_{HVAC}, kW), indoor air temperature (T_{air}, °C), indoor radiant temperature (T_{rad}, °C) and RH (%) as an output. Fourth, day-ahead pricing information was used to create the day-ahead real-time pricing (DARTP) model required for the optimization. Day-ahead energy prices (€/MWh) for the region of central–northern Italy were used as the main component for the formulation of the energy pricing scheme used in the optimization. Additional costs related to transmission/distribution, as well as other costs and taxes, were included to define the final energy pricing profile. Fifth, a GA was constructed to optimize the objective function composed of the daily sum of the hourly cost of energy and the daily average of hourly PMV values for the working hours of the building, specifically from 9:00am to 6:00pm. In the developed GA optimization scheme, HVAC temperature set points were used as the discrete decision variables subject to upper and lower boundaries, which differed between the heating and cooling seasons. Finally, the simulation of the validated building thermal model was executed in an iterative process, using the set points selected by the GA, until convergence criteria were met. Output values of HVAC power simulation, indoor air temperature, radiative temperature and RH were used to evaluate energy cost and the PMV at each iteration.

4.2. GA optimization model

The generic objective function of the GA optimization process is given by the following equation:

$$[min]f\left({T_s}_{i=1}^{J}\right) = w_c \times \frac{\sum_{i=1}^{J} C_i \times P_i}{\sum_{i=1}^{J} C_i \times {P_i}_{baseline}} + w_{pmv} \times \frac{\sum_{i=1}^{J} |PMV_i|}{PMV_{max} \times J} \quad [4.1]$$

subject to $|PMV_i| \leq 1$.

P_i is the HVAC power obtained as an output by the simulation of the building's thermal model and varies according to the building load and temperature set points $({T_s}_{i=1}^{J})$.

PMV_i varies from –3 (cold) to +3 (hot) with zero being the optimum neutral value where internal heat production is equal to the loss of heat to the environment. PMV is calculated according to Standard ISO 7730:2005 based on the following parameters:

– metabolic (M) rate (W/m^2);

– effective mechanical power (W) (W/m^2);

– clothing insulation (I_{cl}) (m^{2}K/W);

– air temperature (T_{air}) (°C);

– mean radiant temperature (T_r) (°C);

– relative air velocity (V_{air}) (m/s).

For the calculation of the PMV hourly values (PMV_i), air temperature, radiant temperature and RH were obtained as an output from the simulation of the building, while certain parameters such as M, W, f_{cl} and p_a were considered to be constant. In the developed approach, the normalized daily average of the PMV hourly absolute values was used to search for optimal near-zero, positive or negative values. Furthermore, the actual values of the PMV were also assessed to reject solutions associated with extreme changes in thermal comfort from one hour to another. This is also prevented based on standards' constraints for set point temperatures drift as explained later in this chapter.

The GA that was developed to optimize the objective function as expressed in equation [4.1], was based on the chromosomes of 24 discrete values (genes) corresponding to the temperature set points of the HVAC for hours 1–24 of the day. Chromosome values were subjected to upper and lower constraints depending on the season of the year. In the heating season, genes $T_{s_{i_h=8}}^{18}$ during the working hours of the building (9:00am to 6:00pm), had a lower boundary of 18°C and an upper boundary of 24°C. In the cooling season, genes $T_{s_{i_c=8}}^{18}$ during the working hours of the building, had a low boundary of 20°C and an upper boundary of 26°C. This is mathematically expressed in the following constraints:

$$18 \leq T_{s_{i_h=9}}^{18} \leq 24$$

$$20 \leq T_{s_{i_c=9}}^{18} \leq 26$$

For two hours prior to the working hours of the building and two hours after, the following constraints were applied to consider preheating and the impact of the extended operation of the HVAC system:

$$17 \leq T_{s_{i_h=7}}^{8} \leq 24$$

$$17 \leq T_{s_{i_h=19}}^{20} \leq 24$$

For two hours prior to the working hours of the building and two hours after, the following constraints were applied to consider precooling and the impact of the extended operation of the HVAC system:

$$20 \leq T_{s}{}_{i_c=7}^{8} \leq 27$$

$$20 \leq T_{s}{}_{i_c=19}^{20} \leq 27$$

4.3. Model of energy cost

According to the utility bills of the Leaf Community in 2018, the average unit cost of energy varied monthly between 0.1507 and 0.1749 €/kWh, as shown in Figure 4.2. Furthermore, it is evident from the graph that the energy consumed outside the peak hours is significant and equal to 35.8%. The two-zone (peak/off-peak) time of use (ToU) pricing scheme, however, offers low incentives for managing loads during daytime, mainly as a consequence of monthly peak power charges.

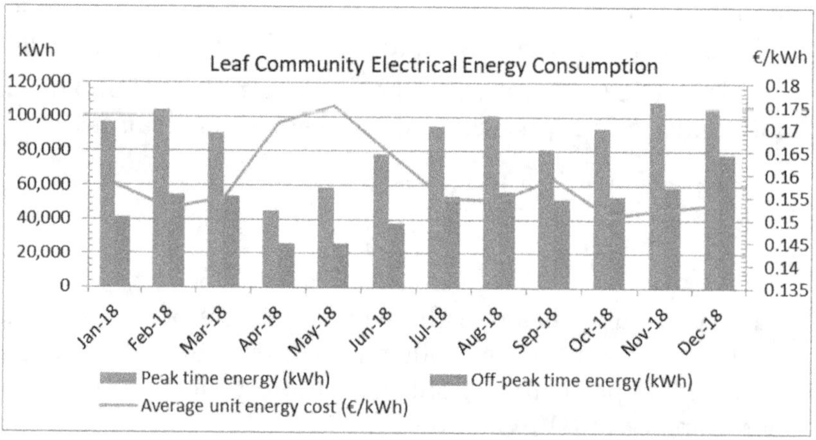

Figure 4.2. *Leaf Community electrical energy consumption and unit cost of energy in 2018. For a color version of this figure, see www.iste.co.uk/kolokotsa/smartbuildings.zip*

The cost of energy model for the Leaf Community was developed in Matlab as described below. Based on current charges related to energy consumption, as identified through energy bills for 2018, basic components were adjusted to incorporate day-ahead hourly price fluctuations in a DARTP scheme and to formulate the case study for dynamic HVAC energy management. Overall, the developed hourly pricing scheme contains costs related to the consumption of energy, maximum power, grid services, taxes and levies. Due to the fact that, in the current pricing scheme, a high share of the costs are determined by fixed charges, these costs were allocated a dynamic parameter to account for network flexibility and stability. The mathematical model of the energy cost is presented in the following equations:

$$C_T = C_{E,T} \times (1 + IVA) \quad [4.2]$$

$$C_{E,T} = C_S + C_N + C_{EDD} \quad [4.3]$$

$$C_S = \sum_{h=0}^{J} E_{hvac,h} \times (DA_h + C_{S,F}) \quad [4.4]$$

$$C_N = C_F + C_{Pmax} + C_{AT} + C_{A-UC} \quad [4.5]$$

$$C_F + C_{AT} + C_{A-UC} = \sum_{h=0}^{J} E_{hvac,h} \times DA_{N,h} \times C_{FAA} \quad [4.6]$$

$$C_{Pmax} = \max(P_{hvac,h}) \times C_{Pmax,F} \quad [4.7]$$

$$C_{EDD} = \sum_{h=0}^{J} E_{hvac,h} \times DA_{N,h} \times C_{EDH} \quad [4.8]$$

4.4. Results and discussion

The generic GA optimization scheme analyzed in the previous section, was applied to include working hours (9:00am to 6:00pm) plus two hours before (7:00–9:00am) and two hours after (6:00–8:00pm). This is considered essential in order to study the effects of preheating and precooling of the building and to maintain internal conditions at comfortable limits for some time after working hours, to account for the fact that some people may still occupy the building. Optimization was conducted for the main industrial thermal zone of the building, which occupies a total area of 1,327 m^2 and a height of 8 m, surrounded by various other spaces including offices, meeting rooms and other facilities over two floors. Following a number of trials, the population size of the GA was set to 50, the crossover fraction was set to 0.8 and the maximum number of iterations was set to 4,600, in order to examine a wide range of different solutions. All solutions obtained during the optimization are stored and the top solutions are filtered in order to assess the set point patterns associated with optimum levels of energy and cost savings, as well as compliance with well-established standards of thermal comfort and temperature set point drift. The approach is designed to evaluate energy cost on a 24-hour time frame. Representative results for four winter days, two days for autumn, one day for summer and one day for spring, are presented to account for different seasonal climatic conditions, heating and cooling modes, as well as DA pricing profiles.

4.4.1. *Scenario 1: January 25, 2018 (winter)*

Results from the GA HVAC optimization on January 25, 2018 are presented in Figure 4.3. In the optimized scenario of this case, set points during working hours were selected, on an hourly basis, to be between 18°C and 22°C, as shown in Figure 4.3(a), and the energy of the HVAC was reduced from 759.88 to 570.13 kWh, a reduction of 25% (Figure 4.3(c)). Energy cost (Figure 4.3(d)) was decreased from €159.3 to €121.03, which is equal to a saving of 25.0%. The HVAC power was kept lower in the optimized scenario, except between 6:00 and 8:00pm. At the baseline scenario, the PMV varied from −0.36 to 0.13, and, at the optimized scenario, from −0.78 to −0.05 (Figure 4.3(b)). The average

PMV varied from −0.14 to −0.38, which corresponds to a PPD increase from 6.28% to 9.13% between the baseline and optimized scenario. In this case, a trade-off between thermal comfort and energy consumption was found to be associated with significant cost savings at times of high pricing rates but also in low pricing zones.

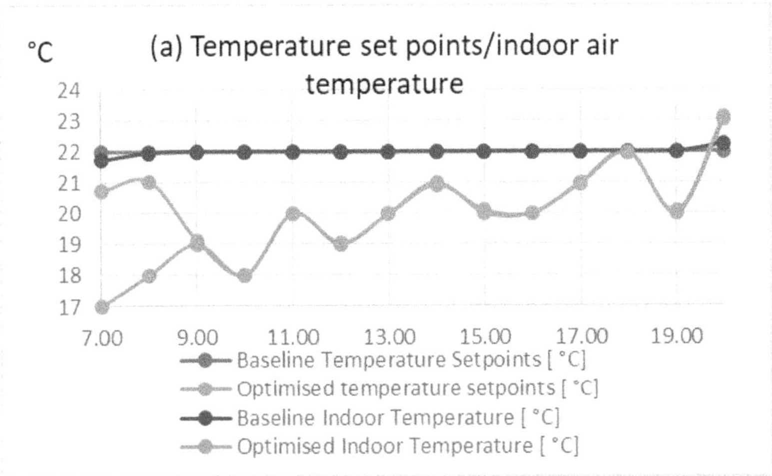

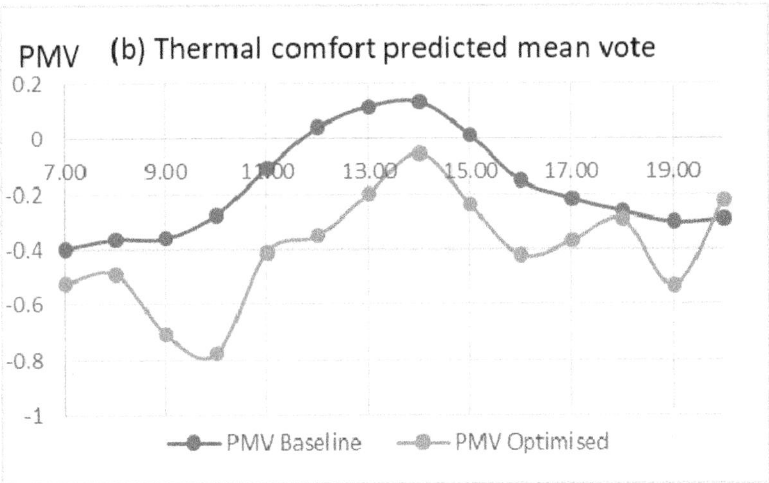

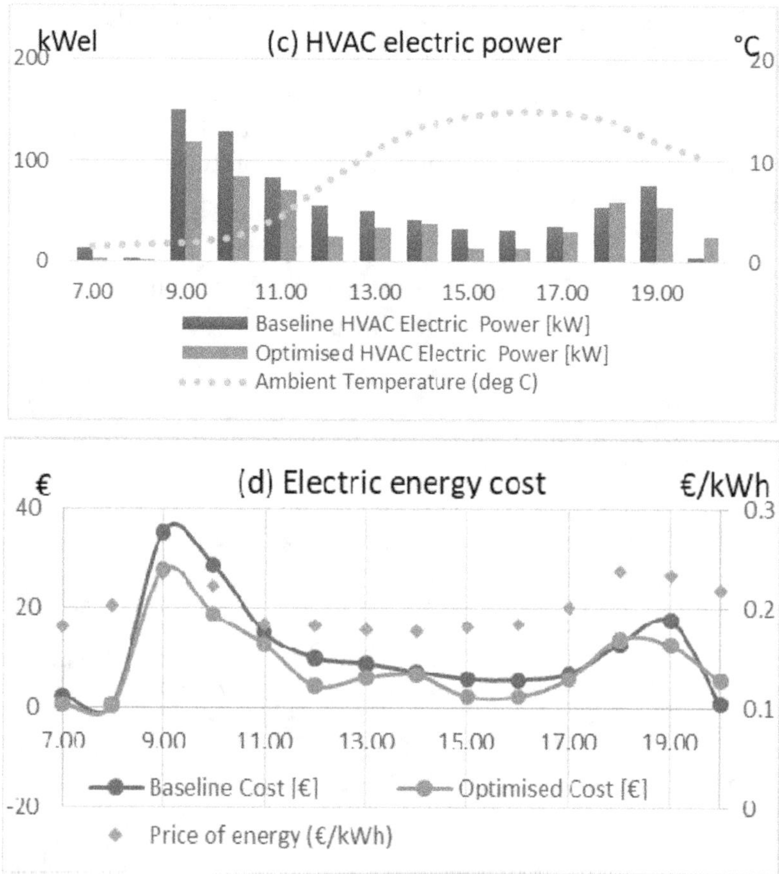

Figure 4.3. *GA HVAC optimization results for January 25, 2018 (winter). For a color version of this figure, see www.iste.co.uk/kolokotsa/smartbuildings.zip*

4.4.2. *Scenario 2: March 27, 2018 (spring)*

Results from the GA HVAC optimization on March 27, 2018 are presented in Figure 4.4. In the optimized scenario of this case, set points were dynamically altered between 19°C and 22°C within working hours (Figure 4.4(a)), and the energy of the HVAC was reduced from 610.91 to 463.43 kWh (Figure 4.4(c)), a reduction of 24.1%. Energy cost (Figure 4.4(d)) was decreased from €121.02 to €94.05, which is equal to a saving of 22.3%. The HVAC power was lower in the optimized scenario, except between the hours of 10:00am and 8:00pm (outside working hours). At the baseline scenario, the PMV varied from –0.39 to –0.02, while, at the optimized scenario, PMV varied from –0.65 to –0.18 (Figure 4.4(b)). The average PMV varied from –0.2 to –0.41, which corresponds to a PPD increase from 6.47% to 9.28% between the baseline and optimized scenario. Similarly, in this case, cost savings occurred mostly early in the morning and late in the evening when prices were relatively high. Some savings were also observed around 12:00–1:00pm just before prices dropped to the lowest level of that particular day.

4.4.3. *Scenario 3: August 15, 2018 (summer)*

Results from the GA HVAC optimization on August 15, 2018 are presented in Figure 4.5. In the optimized scenario, temperature set points varied, on an hourly basis, from 26°C to 24°C, within working hours (Figure 4.5(a)), whereas the energy of the HVAC (Figure 4.5(c)) was reduced from 1,447.83 to 1,175.93 kWh, a reduction of 18.8%. Energy cost (Figure 4.5(d)) was decreased from €295.26 to €238.57, which is equal to a saving of 19.2%. The mean PMV for working hours improved from –0.2 to –0.007, and the PPD was decreased from 6.42% to 5.03%. In the optimized solution, HVAC power was lower during all hours except 2:00, 5:00 and 6:00pm, following the set points changing from higher to lower values down to 24°C. In this scenario, cost savings occurred throughout the day, and were more evident during hours of high prices compared to neighboring regions. The PMV was improved as the fixed cooling set point caused thermal discomfort and unnecessary high energy consumption levels for most hours during the day (Figure 4.5(b)).

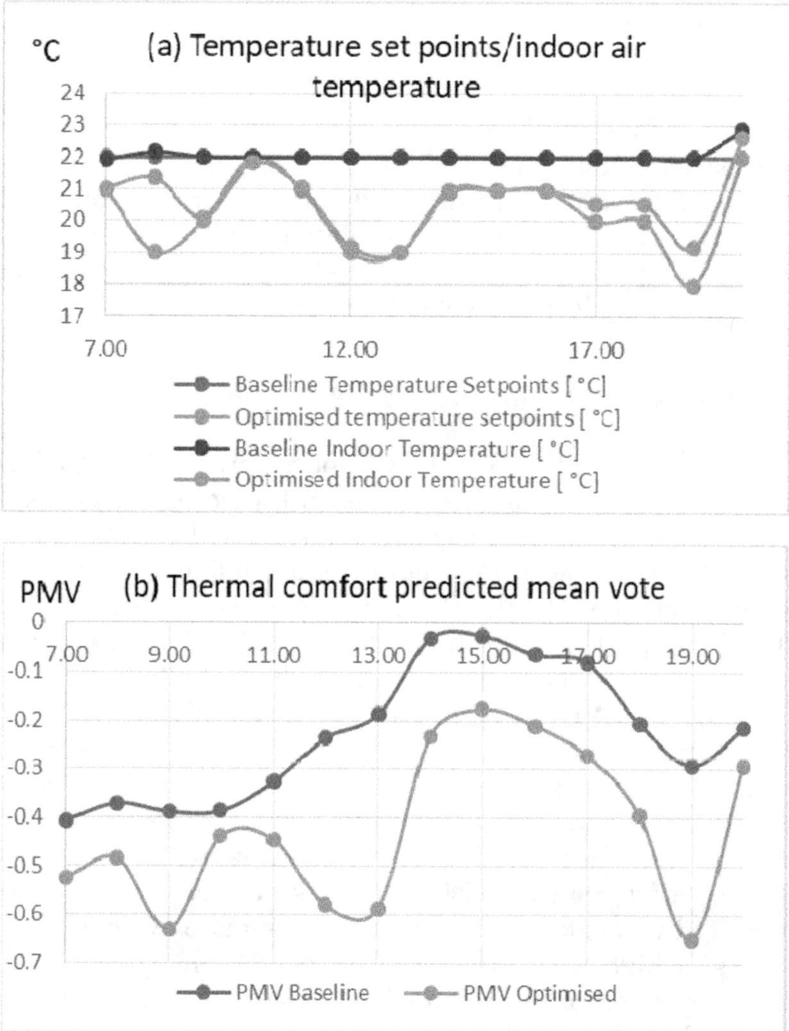
°C
(a) Temperature set points/indoor air temperature
24
23
22
21
20
19
18
17
7.00
12.00
17.00
Baseline Temperature Setpoints [°C]
Optimised temperature setpoints [°C]
Baseline Indoor Temperature [°C]
Optimised Indoor Temperature [°C]
PMV
(b) Thermal comfort predicted mean vote
0
-0.1
-0.2
-0.3
-0.4
-0.5
-0.6
-0.7
7.00
9.00
11.00
13.00
15.00
17.00
19.00
PMV Baseline
PMV Optimised

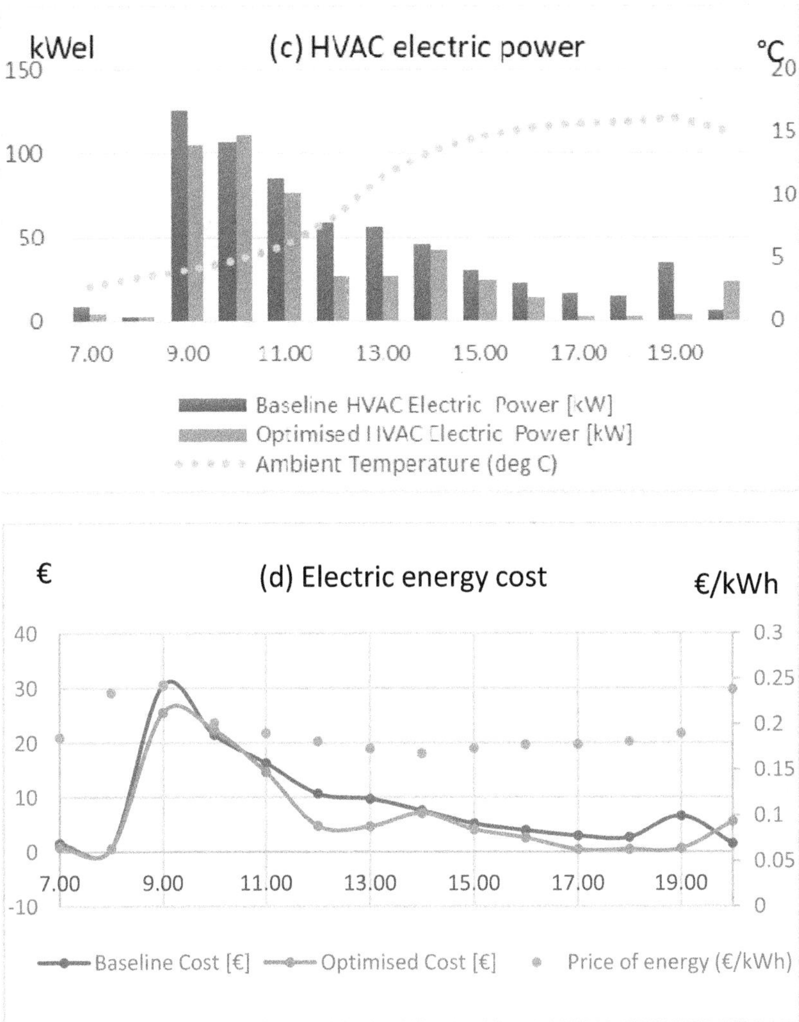

Figure 4.4. *GA HVAC optimization results for March 27, 2018 (spring). For a color version of this figure, see www.iste.co.uk/kolokotsa/smartbuildings.zip*

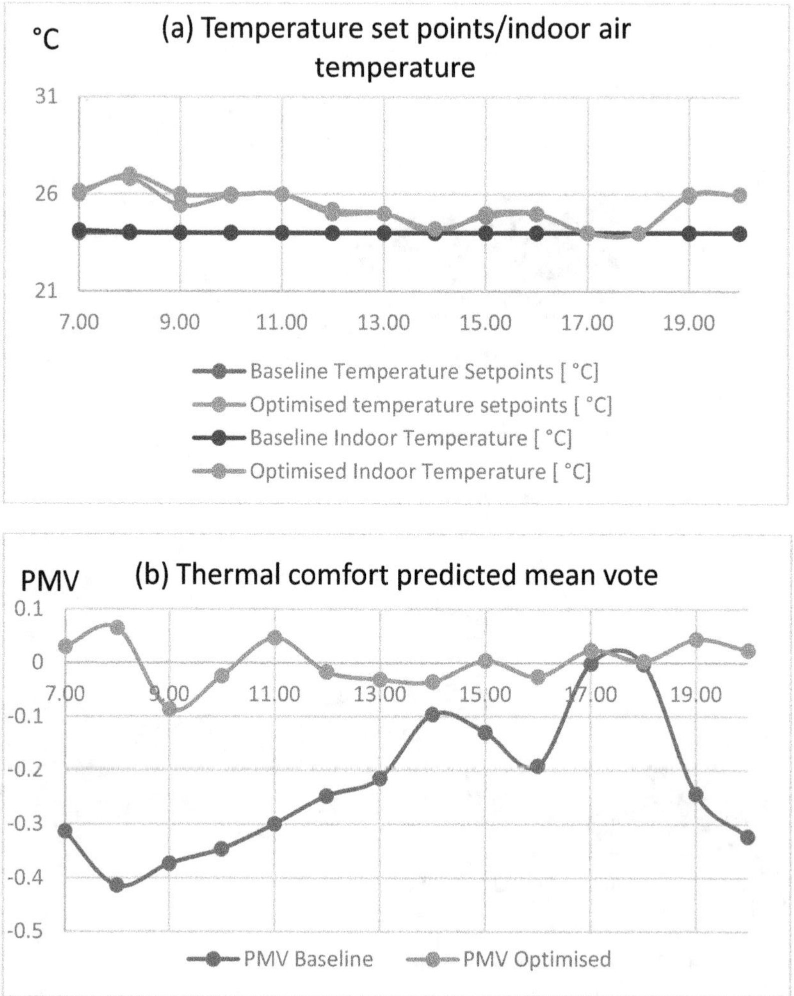
°C
(a) Temperature set points/indoor air temperature
31
26
21
7.00
9.00
11.00
13.00
15.00
17.00
19.00
Baseline Temperature Setpoints [°C]
Optimised temperature setpoints [°C]
Baseline Indoor Temperature [°C]
Optimised Indoor Temperature [°C]
PMV
(b) Thermal comfort predicted mean vote
0.1
0
-0.1
-0.2
-0.3
-0.4
-0.5
7.00
9.00
11.00
13.00
15.00
17.00
19.00
PMV Baseline
PMV Optimised

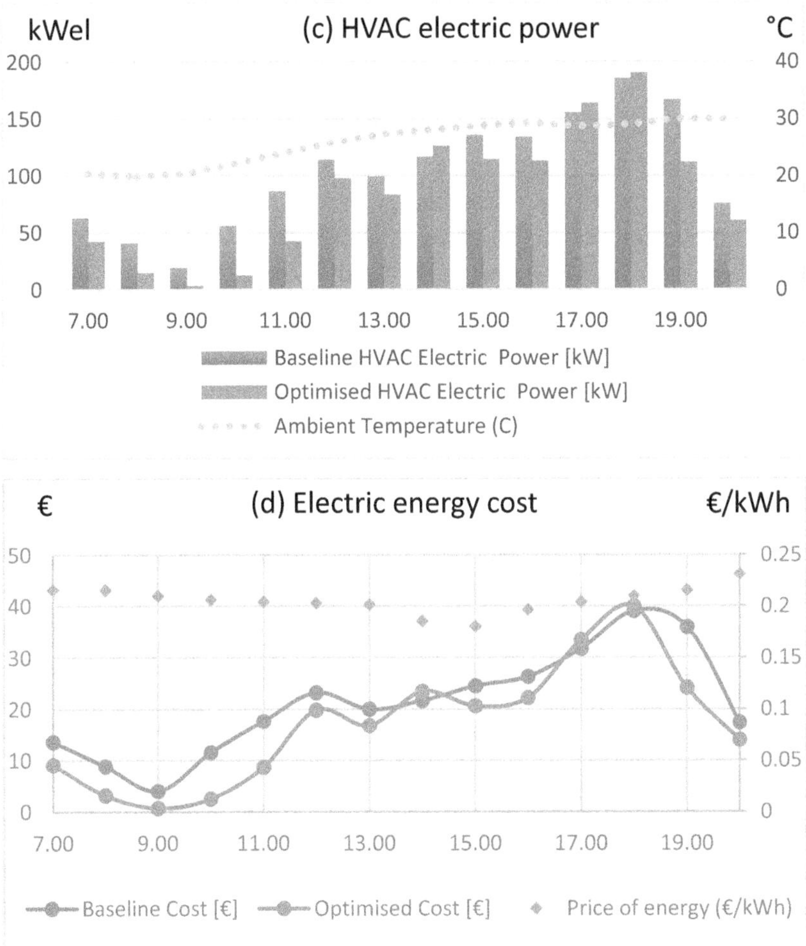

Figure 4.5. *GA HVAC optimization results for August 15, 2018 (summer). For a color version of this figure, see www.iste.co.uk/kolokotsa/smartbuildings.zip*

4.4.4. *Scenario 4: September 10, 2018 (autumn)*

Results from the GA HVAC optimization on September 10, 2018 are presented in Figure 4.6. In this case, the energy of the HVAC (Figure 4.6(c)) was reduced from 1,268.47 to 1,136.29 kWh, a reduction of 10.4%, while energy cost (Figure 4.6(d)) was decreased from €311.59 to €280.68, a reduction of 9.9%.

The slight difference in the mean PMV for working hours from −0.144 to −0.073 corresponds to a PPD decrease of 1.1%. The PMV at the baseline scenario varied from −0.56 up to 0.11, while, in the optimized scenario, PMV varied from −0.39 to 0.06 (Figure 4.6(b)). HVAC power values (Figure 4.6(c)) in the optimized scenario exceeded their respective values in the baseline scenario at times of low prices with respect to the neighboring regions and specifically from 12:00 to 3:00pm. Baseline energy consumption was unnecessarily high during the morning as it coincided with significant negative PMV levels, while efficient performance was observed in the optimized scenario where the set point was kept at the highest level allowed. Indoor temperature (Figure 4.6(a)) deviated from the set point temperature for both the baseline and the optimized scenario between 9:00 and 10:00am. In the optimized scenario, the HVAC energy consumption remained at a low level due to the negative PMV levels during the same time period. Another interesting observation was that the PMV in the baseline scenario significantly increased, over time, during the day, despite the fact that the indoor temperature was kept constant, which was mainly due to the effect of radiative temperature on thermal comfort.

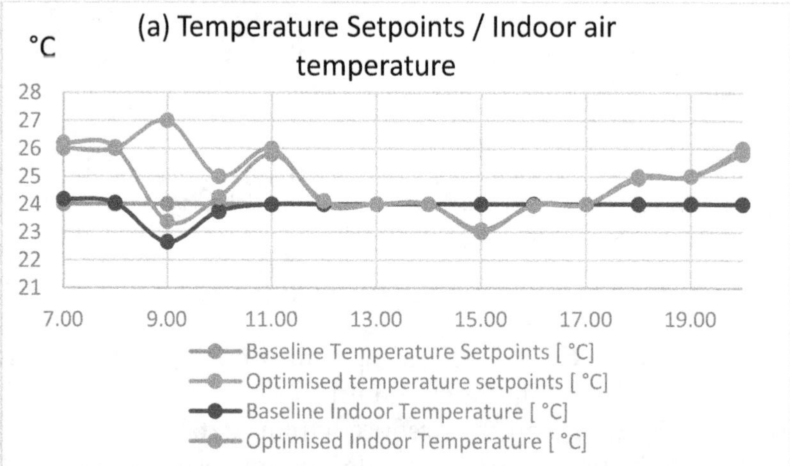

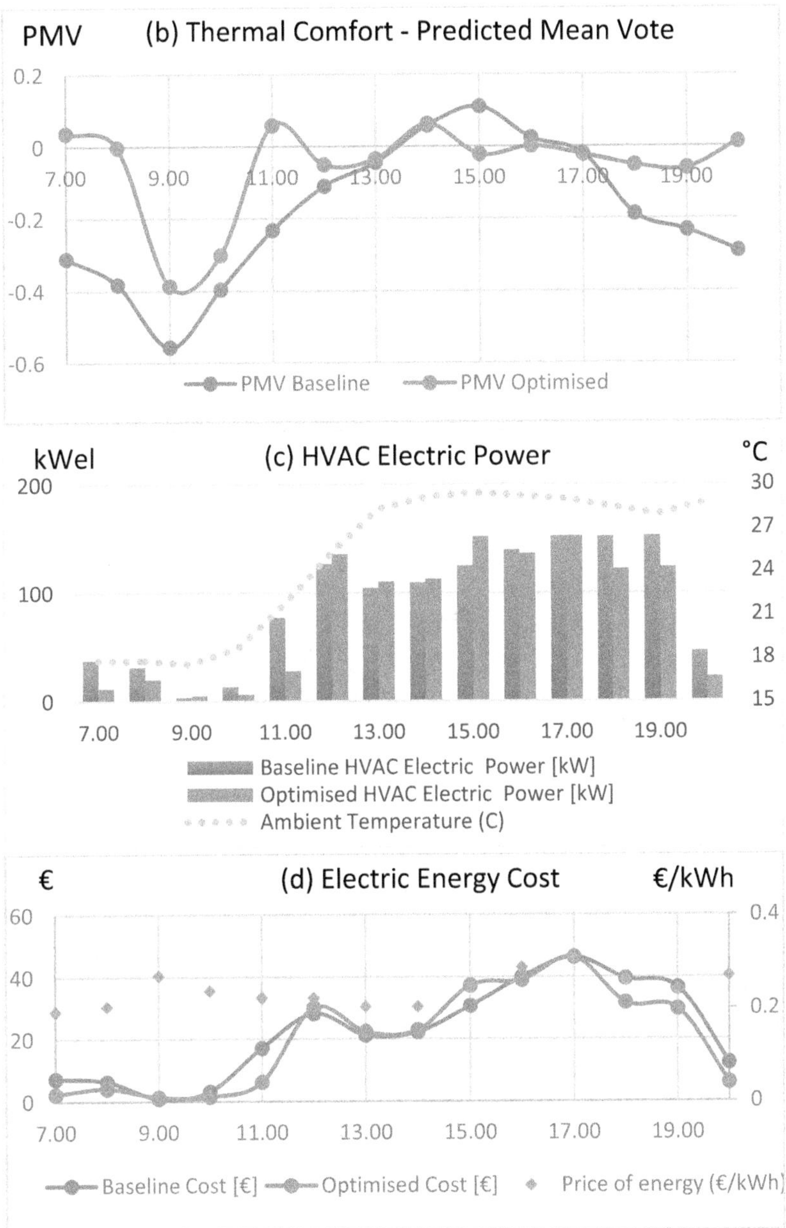

Figure 4.6. *GA HVAC optimization results for September 10, 2018 (autumn). For a color version of this figure, see www.iste.co.uk/kolokotsa/smartbuildings.zip*

4.4.5. *Scenario 5: September 21, 2018 (autumn)*

Results from the GA HVAC optimization on September 21, 2018 are presented in Figure 4.7. In the optimized scenario, set points within working hours fluctuated between 26°C and 24°C (Figure 4.7(a)), while the energy of the HVAC was reduced from 1,248.69 to 1,078.16 kWh (Figure 4.7(c)), a reduction of 13.7%. Energy cost (Figure 4.7(d)) was decreased from €298.07 to €253.79, which is equal to a saving of 14.9%. The mean PMV for working hours improved from –0.172 to –0.056, and the respective PPD was decreased from 6.4% to 5.4%. In this case, the optimized PMV (Figure 4.7(b)) reflected improved thermal conditions, since it oscillated in the region –0.40 to 0.05 in the optimized scenario, while, in the baseline scenario, it varied between –0.56 and 0.05. Energy savings were achieved by keeping the temperature set points at higher levels, while the PMV was at negative levels during the early morning and late afternoon working hours. Slightly higher HVAC power levels (Figure 4.7(c)) were observed between 12:00–3:00pm, coinciding with the lowest energy prices during the day.

4.4.6. *Scenario 6: November 20, 2018 (winter)*

Results from the GA HVAC optimization on November 20, 2018 are presented in Figure 4.8. Temperature set points in the optimized scenario were dynamically controlled from 17°C to 23°C (18–22°C within working hours; Figure 4.8(a)). In the optimized scenario, the energy of the HVAC was reduced from 923.75 to 756.67 kWh, a reduction of 18.1% (Figure 4.8(c)). Energy cost, shown in Figure 4.8(d), was decreased from €217.33 to €177.66, which is equal to a saving of 17.4%. PMV in the optimized scenario varied from –0.50 to –0.10, whereas, in the baseline scenario, it varied between –0.07 and –0.05 (Figure 4.8(b)). The mean PMV for working hours increased from –0.055 to –0.276, and the respective PPD increased by 1.4%. PMV was kept at small negative values and above −0.3 for most hours, except for 2:00 and 3:00pm, where PMV was −0.49 and −0.35, respectively. HVAC power in the optimized scenario was kept at lower levels compared to the baseline for all working hours (except early morning hours).

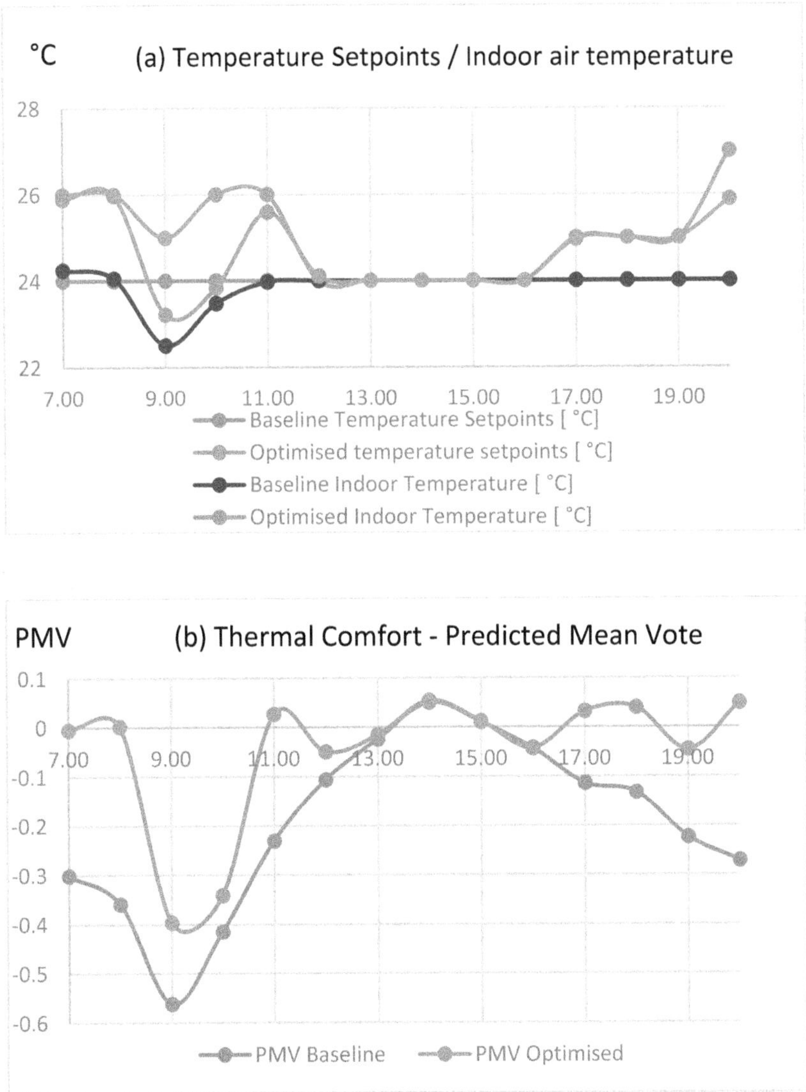

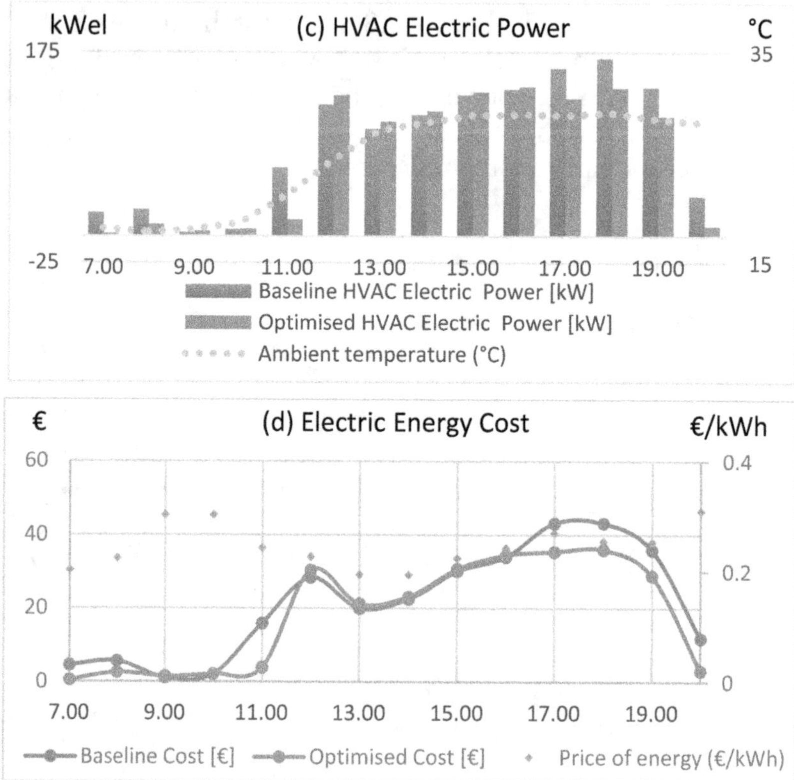

Figure 4.7. *GA HVAC optimization results for September 21, 2018 (autumn). For a color version of this figure, see www.iste.co.uk/kolokotsa/smartbuildings.zip*

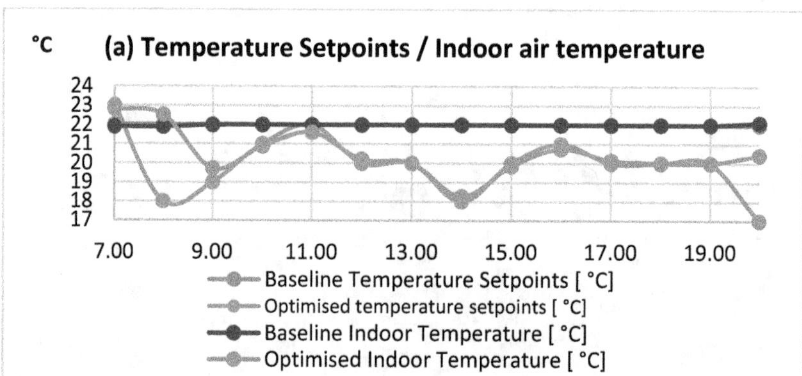

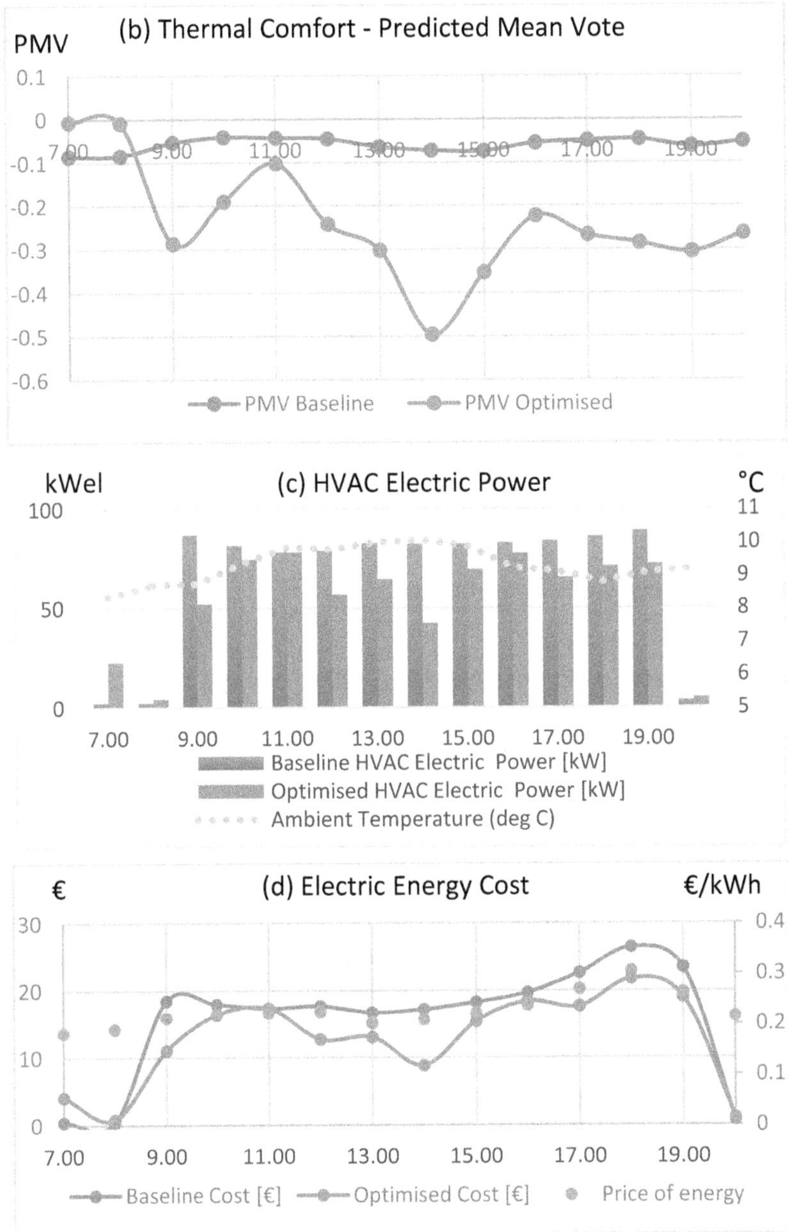

Figure 4.8. *GA HVAC optimization results for November 20, 2018 (winter). For a color version of this figure, see www.iste.co.uk/kolokotsa/smartbuildings.zip*

4.4.7. *Scenario 7: November 22, 2018 (winter)*

Results from the GA HVAC optimization on November 22, 2018 are presented in Figure 4.9. Optimized temperature set points, as presented in Figure 4.9(a), varied from 19°C to 22°C. In the optimized scenario, the energy of the HVAC was reduced from 717.77 to 631.61 kWh, a reduction of 12.0% (Figure 4.9(c)). Energy cost (Figure 4.9(d)) was decreased from €179.59 to €159.01, which is equal to a saving of 11.5%. PMV in the optimized scenario (Figure 4.9(b)) varied between –0.25 and –0.01 and, in the baseline scenario, PMV varied between –0.03 and –0.01. The mean PMV for working hours was increased from –0.016 to –0.110, and the respective PPD was increased by 0.4%. Significant energy savings occurred during hours of high prices, at 9:00am, 1:00pm and 3:00–6:00pm, while the PMV was kept at small negative levels down to –0.25.

4.4.8. *Scenario 8: November 25, 2018 (winter)*

Optimization results for HVAC optimization on November 25, 2018 are presented in Figure 4.10. Temperature set points varied from 18°C to 22°C in the optimized solution (Figure 4.10(a)). In this scenario, HVAC energy consumption (Figure 4.10(c)) was reduced from 944.85 to 776.17 kWh, a decrease of 17.9% compared to the baseline. Daily cost (Figure 4.10(d)) was reduced from €199.52 to €164.26, a reduction of 17.7%. The mean PMV was decreased from –0.244 in the baseline scenario to –0.478, which is equivalent to a PPD increase of 4.2% (Figure 4.10(b)). HVAC consumption in the optimized scenario exceeded the baseline levels slightly at hours 8:00am, and 5:00 and 6:00pm. A compromise was reached by maintaining comfort within tolerable levels while maximizing cost savings.

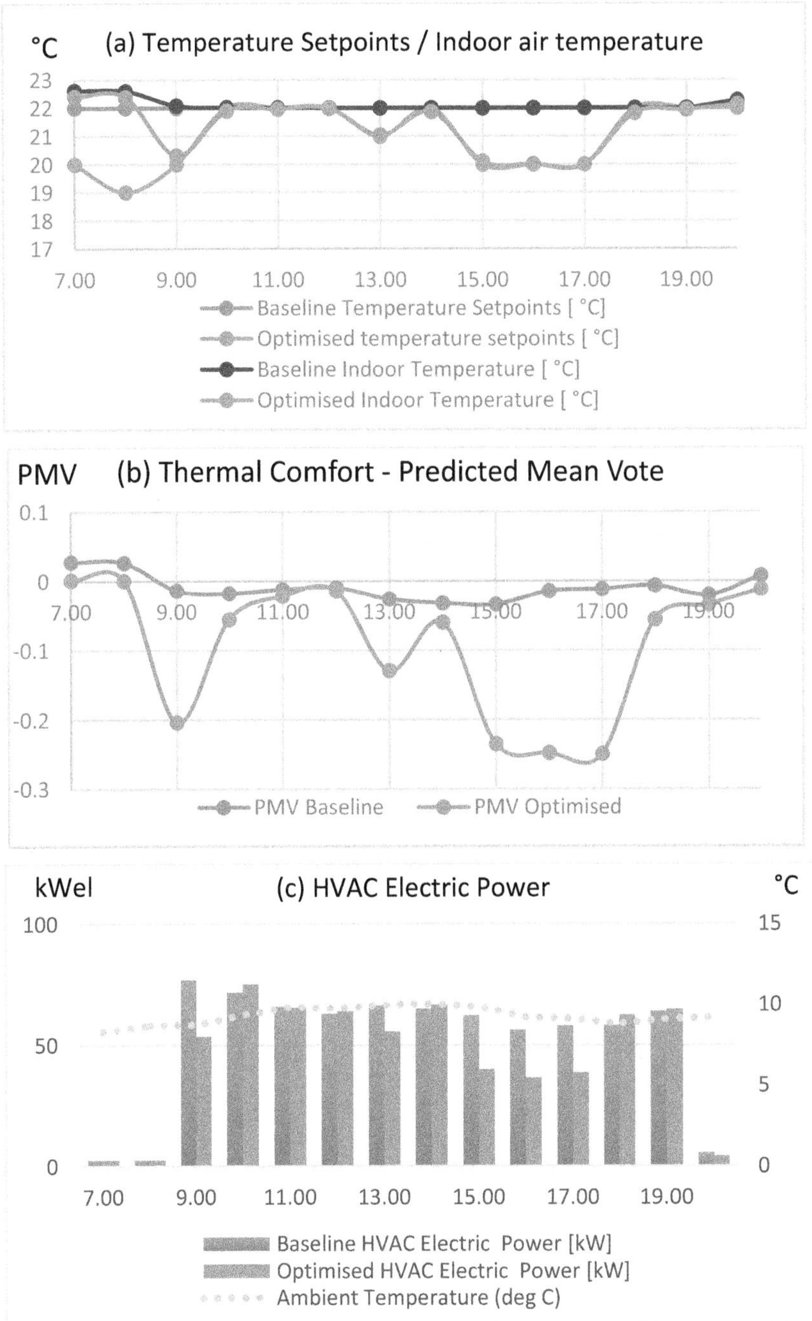
°C
(a) Temperature Setpoints / Indoor air temperature
23
22
21
20
19
18
17
7.00
9.00
11.00
13.00
15.00
17.00
19.00
Baseline Temperature Setpoints [°C]
Optimised temperature setpoints [°C]
Baseline Indoor Temperature [°C]
Optimised Indoor Temperature [°C]
PMV
(b) Thermal Comfort - Predicted Mean Vote
0.1
0
-0.1
-0.2
-0.3
7.00
9.00
11.00
13.00
15.00
17.00
19.00
PMV Baseline
PMV Optimised
kWel
(c) HVAC Electric Power
°C
100
50
0
15
10
5
0
7.00
9.00
11.00
13.00
15.00
17.00
19.00
Baseline HVAC Electric Power [kW]
Optimised HVAC Electric Power [kW]
Ambient Temperature (deg C)

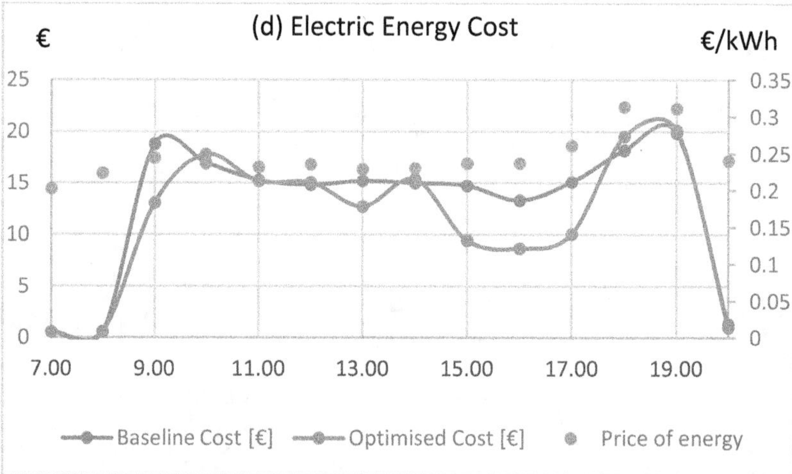

Figure 4.9. *GA HVAC optimization results for November 22, 2018 (winter). For a color version of this figure, see www.iste.co.uk/kolokotsa/smartbuildings.zip*

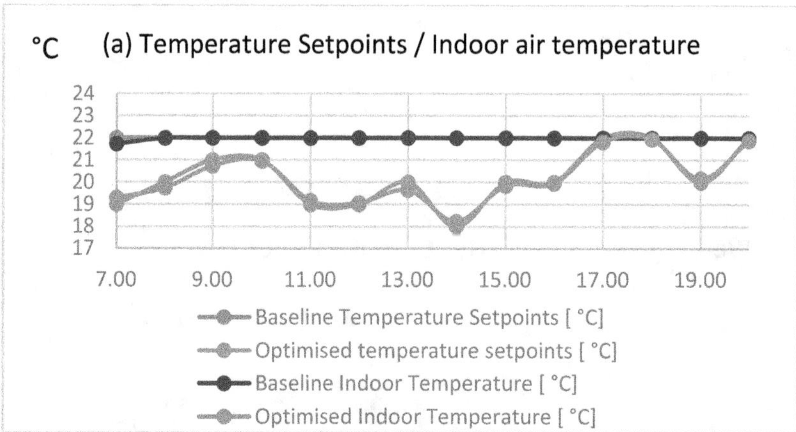

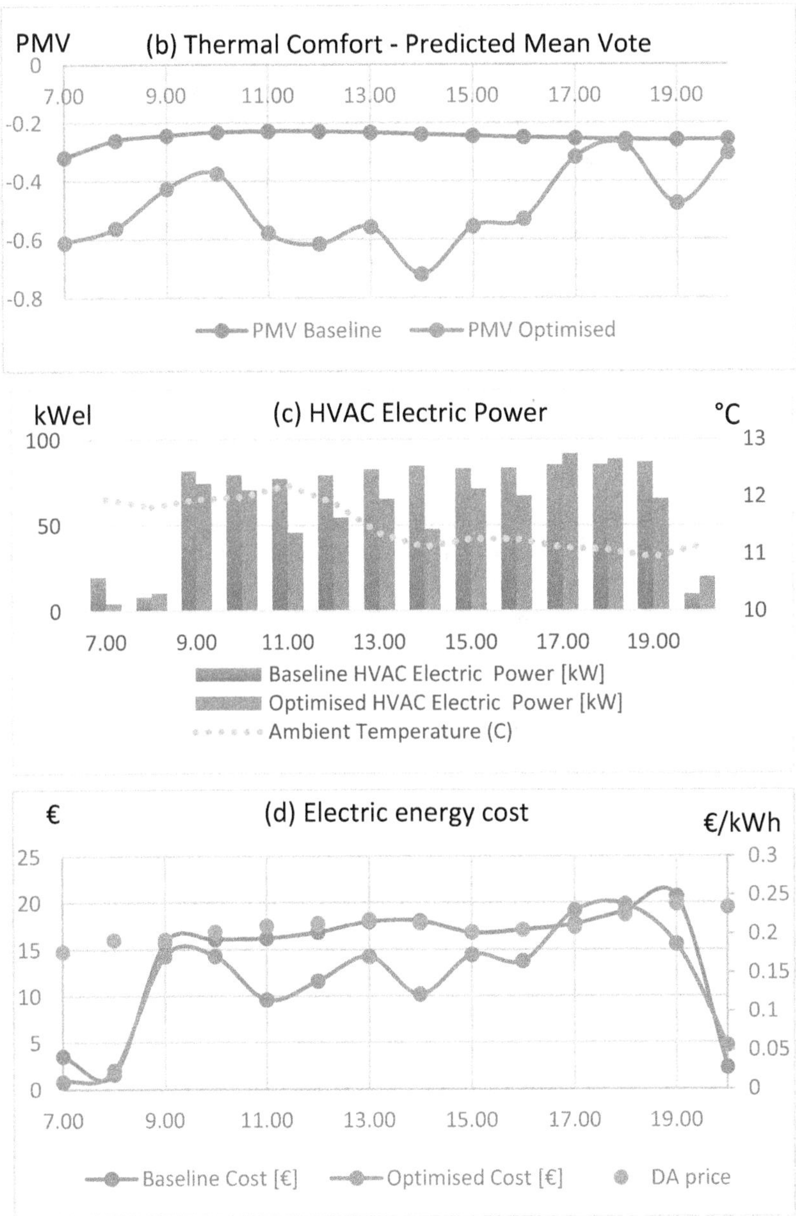

Figure 4.10. *GA HVAC optimization results for November 25, 2018 (winter). For a color version of this figure, see www.iste.co.uk/kolokotsa/smartbuildings.zip*

4.5. Conclusion and future steps

The developed approach provides an optimization assessment framework for HVAC energy management, in day-ahead real-time pricing DR programs. In this framework, a GA-based approach was developed to investigate DR implementation for a near-zero energy industrial building located in the region of Marche in Italy. Results indicate that there is significant potential for energy and cost savings by controlling indoor conditions with acceptable levels of thermal comfort, as evaluated according to PMV. Several scenarios were analyzed to demonstrate a realistic potential cost of energy reductions in the range of between 9.9% and 25%, whereas the reduction of HVAC energy consumption varied between 10.4% and 25%. Presented solutions within established standard requirements for indoor comfort and indoor temperature drift rate were selected for evaluation from a wide range of available solutions. The proposed DR approach is applicable in a wide range of building energy optimization assessment schemes due to the fact that it deploys temperature set point levels for HVAC control. It can be applied to establish optimal control of thermal zones in buildings of various uses and sizes, controlled by single or distributed thermostatic controls. Results demonstrate that there is an unexplored potential for HVAC dynamic control associated with DR RTP schemes, which could be intelligently embedded in the operation of such systems in the years to come. The computational cost of the proposed approach was significant, as, at this stage of the research, a high number of iterations (4,600) were performed to ensure the search was as extensive and as deep as possible. However, based on the results obtained, there is great potential for reducing the time for GA convergence, since satisfactory near-optimal results were, in most cases, obtained in the first day of the run (total average time for a conventional personal computer was approximately two days). Furthermore, a careful adjustment of optimization parameters and constraints combined with weather predictions, along with the evolution of computer resources and microprocessor capabilities, can make the proposed approach deployable in real time in the near future. In addition, future research could involve the investigation of a typology of HVAC near-optimal set point configurations, in response to patterns in ambient conditions. Experimental research could entail the testing of optimal set point patterns in real conditions as the next step towards the actual implementation of the developed methodology.

5

Smart Grid/Community Load Shifting GA Optimization Based on Day-ahead ANN Power Predictions

Preparation for the transition from conventional power grids to next generation, so-called, "smart" grids is a worldwide trend nowadays. The goal for stakeholders in the domains of operations, generation, transmission, distribution and service provision (Greer *et al.* 2014) is to offer more, and higher quality, services while improving operational capabilities, flexibility and energy efficiency. In this context, a higher level utilization of smart grid resources is targeted by grid modernization and enhanced, dispersed dynamic measurements at local, regional and wider levels. Various forms of communication equipment and protocols allow smart metering, monitoring and controls in an interoperable unified system, often described as an advanced metering infrastructure (AMI). Several architectures and network topologies have been proposed to accommodate a reliable and efficient exchange of bidirectional flow of energy

---

Chapter written by Nikos KAMPELIS, Elisavet TSEKERI, Denia KOLOKOTSA, Konstantinos KALAITZAKIS, Daniela ISIDORI and Cristina CRISTALLI. This chapter is based upon: "Development of demand response energy management optimization at building and district levels using genetic algorithm and artificial neural network modelling power predictions," *Energies*, 2018.

and information. In Jain *et al.* (2019), consumer demand is prioritized and demand response (DR) data throughput is optimized, enabling a faster reaction.

Smart metering and AMI are widely recognized as a necessity for the reliable and fast exchange of data in smart grids (Rietveld *et al.* 2012). It is expected that nodal analysis of power measurements in the power grid will provide valuable information for utilities to control multidirectional flows of energy and improve dispatching, addressing vulnerabilities and constraints.

In this sense, it is foreseen that a variety of technological solutions will emerge to balance the high volatility and power quality issues of the miscellaneous intermittent loads and renewable energy sources.

On the market side, reforms are required to leverage innovation in services and new business models which will upgrade existing operations. In this context, DR constitutes a variety of services that have transformed the electric grid and energy market operations during the past decades. Significant progress has been made in the United States, where DR programs have been designed and implemented for years, and span across the full range of dispatchable (reliability, economic) and non-dispatchable (time sensitive pricing, time of use (ToU), critical peak pricing (CPP), real-time pricing (RTP)) demand side management options (NERC 2011). Demand side management is a valuable prospect for consumers and utilities – if used properly – for the use of assets to decrease losses in transmission and distribution, as well as to reduce avoidable costs. In this context, DR, along with the demand side management of distributed energy resources (DER), expands the boundaries for near future scientific and technological advances.

In the European Union (EU), the Energy Efficiency Directive (EED), 2012/27/EU, foresees the elimination of barriers for DR in balancing and ancillary services markets (European Parliament and the Council of the European Union and the Council of the European Union 2012). Among the EU member states (MS), considering the progress in DR, Belgium, France, Ireland, and the UK, are in the leading group. Significant steps have also been taken in this direction by Germany, the Nordic countries, the Netherlands and Austria. Generally, DR programs are differentiated: (a) explicitly, i.e. where DR participants transact directly within the energy market; and (b) implicitly, i.e. where participation through a third party is facilitated (Bertoldi *et al.* 2016).

The overall framework of smart grids with regard to DR is presented and analyzed by Siano (2014). Important aspects are defined, and a description of the possibilities created by DR for utilities and customers are analyzed. Load curtailment, shifting energy consumption and using onsite energy generation, thus reducing dependence on the main grid, are the main mechanisms for customers to participate in DR. Customer participation in wholesale markets via intermediaries, such as curtailment service providers (CSP), aggregators, or retail customers (ARC), demand response providers (DRPs), or local distribution companies, is documented in Siano (2014). Moreover, a review of DR and smart grids with respect to the potential benefits and enabling technologies is provided. Considering system operation, contingency issues can be dealt with through DR implementation, resulting in a reduction of electrical consumption at critical hours, and avoiding serious impacts due to failure of power services provision. Considering energy efficiency, it is ascertained that effective management of aggregated loads can lead to a reduction of the overall cost of energy, due to the reduction and operating time shortening of conventional power generation equipment. Avoiding network upgrades at the local level, or postponing investments in new capacity, reserves or peaking units at the system level, is another important potential benefit linked to high level implementation of DR. Modeling of incentive-based DR, focusing on interruptible/curtailable service and capacity market programs, is investigated by Aalami *et al.* (2010). Price elasticity of demand and a customer benefit function are used to develop an economic model. Several scenarios are simulated and evaluated according to different strategies, improvement of the load curve (peak reduction, load factor and peak to valley), the benefit to customers and reduction of energy consumption.

Wholesale electricity market design considerations, with regard to major challenges, aimed at increasing renewable energy penetration are explored in Kusiak *et al.* (2011). Various dynamic energy pricing models have been proposed to compensate for market uncertainty and risks (Triki and Violi 2009; Ghazvini *et al.* 2017). A residential DR based on adaptive consumption pricing is proposed by Haider *et al.* (2016), allowing utilities to manage aggregate load and customers to lower their energy consumption. The proposed pricing scheme adapts energy costs to customers' consumption levels, thus encouraging active enrolment in the DR program. Cost and comfort optimization of load scheduling under different pricing schemes has been investigated using various techniques, including linear optimization, convex optimization, particle swarm optimization (PSO) and mixed integer nonlinear programming (MINLP) (Tsui and Chan 2012). Furthermore, technology readiness, opportunities and requirements for the deployment of DR in buildings and blocks of buildings are addressed by Crosbie *et al.* (2017, 2018).

On the other hand, buildings worldwide are responsible for over 40% of total energy consumption, gas emissions and global warming (Attia *et al.* 2018). The role of smart grids for near- and zero-energy building communities is investigated by researchers to test new approaches, identify critical aspects and tackle challenges the emerge when dealing with design and operational problems (Kolokotsa 2016; Karlessi *et al.* 2017). On the demand side, a wide variety of developed scientific tools influence the dynamics of advances in energy performance and energy management in buildings (Kampelis *et al.* 2006; Diakaki *et al.* 2008, 2010; Kolokotsa *et al.* 2009). Such tools are embedded in data monitoring applications, such as innovative web-based energy management platforms (Kolokotsa *et al.* 2016; Gobakis *et al.* 2017), to enable improved analysis, decision-making and dynamic controls. Moving from building energy management systems (BEMSs) (Zhou *et al.* 2016; Shakeri *et al.* 2017) to district energy management systems (DEMSs) (Fanti *et al.* 2015) entails the dynamic exchange and hierarchical processing of data streams between various components and systems, as in the Internet of Things (IoT) paradigm (Brusco *et al.* 2017; Marinakis and Doukas 2018). Various techniques and tools have been investigated for dealing with challenges in various fields pertaining to smart grids: smart metering data analysis and dynamic processing (Lu *et al.* 2012); power demand forecasting (Baliyan *et al.* 2015; Raza and Khosravi 2015); DER management optimization (Shinkhede and Joshi 2014); user engagement (Goulden *et al.* 2014), etc.

DR is a fundamental aspect of the smart grid concept as it refers to the necessary open and transparent market framework linking energy costs to actual grid operations. DR allows consumers to directly or indirectly participate in the markets where energy is being exchanged. One of the main challenges for engaging in DR is associated with the initial assessment of the potential rewards and risks under a given pricing scheme. In this chapter, a genetic algorithm (GA) optimization model is proposed, using artificial neural network (ANN) power predictions for day-ahead energy management at building and district level. Individual building and building group analyses are conducted to evaluate ANN predictions and GA-generated solutions. ANN-based short-term electric power forecasting is exploited in predicting day-ahead demand and forms a baseline scenario. GA optimization is conducted to provide balanced load shifting and cost of energy solutions, based on alternative pricing schemes. Results demonstrate the effectiveness of this approach for assessing DR load shifting options based on ToU and day-ahead real-time pricing (DARTP) schemes. Through the analysis of the results, the practical benefits and limitations of the proposed approach are addressed.

The chapter is organized as follows. In section 5.1, the infrastructure and the applied methodology are presented. The proposed day-ahead GA approach for the cost of energy and load shifting optimization based on ANN hourly power predictions is analyzed in section 5.2. Results of ANN power predictions and GA load shifting optimization, based on a ToU pricing scheme, are presented in section 5.3, whereas results of ANN power predictions and GA load shifting optimization, based on a DARTP scheme, are provided in section 5.4. Further discussion on ANN power predictions and GA-based obtained solutions are provided in section 5.5. Finally, in section 5.6, conclusions and recommendations for future work are summarized.

5.1. Infrastructure and methods

The proposed novel approach was developed and tested on the basis of data available from the MyLeaf platform which monitors and controls the Leaf Community buildings. The buildings in the Leaf Community are highly thermally insulated and are equipped with automations for controlling the heating, ventilation and air conditioning (HVAC) systems, as well as the natural and artificial lighting by means of adjustable external louvers and luminance sensors. The primary annual energy consumption for the Leaf Lab is rated at 35.4 kWh/m^2 (including the PV power production and subtracting industrial consumption) (Kampelis *et al.* 2017) based on year-round measurements, while the L6 is estimated at 46.85 kWh/m^2. Table 5.1 summarizes the basic components of the building envelopes and systems installed at the Leaf Community buildings under consideration. A detailed description of the Leaf Community is provided in section 5.2.

Pilot case studies	Sky windows	Automatic shading	Illuminance/ presence light controls	LED	Ground water heat pumps	biPV	Thermal storage	Electrical Storage
L2: Summa – Offices/Warehouse (1,037 m^2)			•	•	•	•		•
L4: Leaf Lab – Industrial (6,000 m^2)	•	•	•	•	•	•	•	•
L5: Kite Lab – Offices, Laboratories (3,514 m^2)	•		•	•	•	•		•

Table 5.1. *Pilot buildings in the Leaf Community*

The methodology developed comprises several steps, as shown in Figure 5.1.

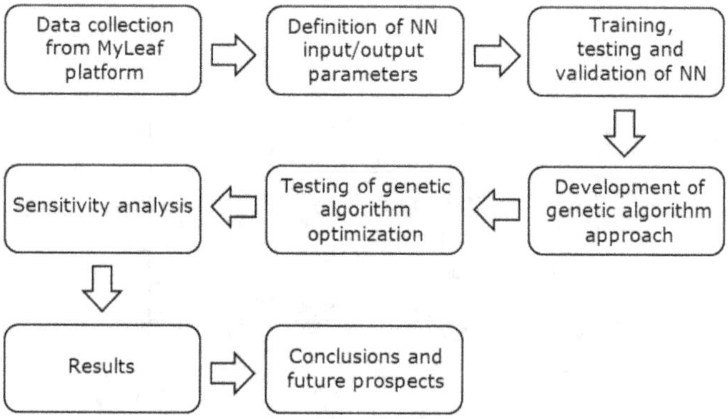

Figure 5.1. *Methodological framework*

1) *Collection of data*: all data from measuring equipment, sensors and actuators in the Leaf Community are collected, organized and made remotely available through the MyLeaf platform (Song *et al*. 2019). In this case, the MyLeaf platform is used to collect data of ambient temperature, irradiance and power demand of the buildings considered in the analysis.

2) *Development and testing of ANN models*: ANN models are developed and exploited to perform day-ahead predictions of power consumption Matlab. For the 24-hour ahead prediction of power consumption, the day of the week, the time, irradiance and the external temperature are used as inputs, while the 24-hour ahead net electrical power is used as the target. Trials of various combinations for the ANN model parameterization are performed, considering the structure, the algorithm, the number of hidden layers and

the delays. A Lavemberg–Marquardt algorithm was deployed in a nonlinear autoregressive ANN structure with exogenous input (NARX), with three hidden layers and a delay of one.

3) *GA load shifting approach*: A GA optimization scheme was developed and tested in MATLAB in order to provide alternative solutions for load shifting. The GA optimization scheme is based on the developed mathematical model analyzed in section 5.2. The objective function includes the criteria of energy cost and load shifting. Market information is used to construct the hourly pricing profiles used in the optimization process. Weighting coefficients are applied to both normalized criteria to enable consideration of several alternatives, depending on several priorities and energy management capabilities. Weighting coefficients are used to provide a trade-off between cost and load shift. The role of weighting coefficients is to allow a decision-maker to investigate a set of solutions and obtain those that better match his/her preferences. Preferences differ based on the decision-maker's knowledge and understanding but may also be influenced by other priorities during various time periods. For example, cost savings could be considered to be the "default" priority but, during certain periods, minimization of load shifting could be upgraded to become the dominant factor in the optimization process.

4) *Sensitivity analysis and evaluation of results*: sensitivity analysis is performed by changing the GA parameters, such as crossover, population size, mutation rate and tolerance. Furthermore, since load shifting is related to changes in the operation of building systems (HVAC, lighting, etc.) and operations (industrial, office), it also needs to be minimized in order to avoid significant intervention in the buildings' use. On the other hand, the cost of energy is minimized when load shifting occurs from

hours of high prices to hours of low prices. The solutions are hence evaluated considering the hourly/daily cost of energy and load shifting preferences.

The developed approach is illustrated with the aid of the flowchart in Figure 5.2.

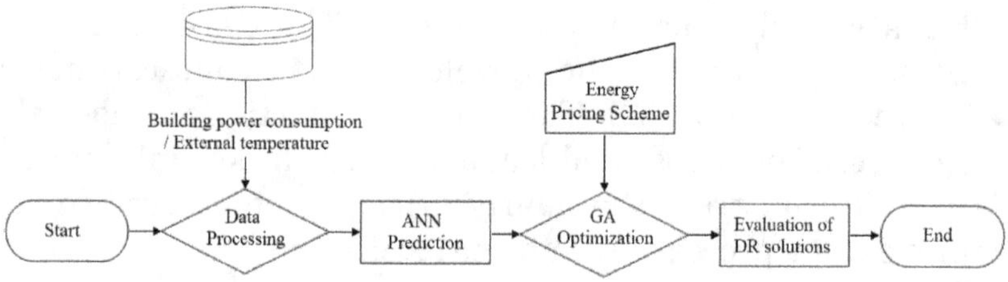

Figure 5.2. *Flowchart of the developed approach*

5.2. Day-ahead GA cost of energy/load shifting optimization based on ANN hourly power predictions

The GA optimization scheme is based on the developed mathematical model presented hereafter. The two criteria, namely the normalized cost of energy and load shifting, form the objective function, as shown in equation [5.1]:

$$f = \min\left(w_1 \frac{Cost_E}{Cost_{E_{max}}} + w_2 \frac{Load_{Shift}}{Load_{Shift\_max}}\right) \quad [5.1]$$

At the building group level, the cost term of the objective function in equation [5.1] is given by equation [5.2].

$$Cost_E = \sum_{b}^{B} Cost_E^b \quad [5.2]$$

where b is used to denote each building that belongs to the group.

The energy cost of each building in equation [5.2] is calculated using equation [5.3]:

$$Cost_E^b = \sum_{h=1}^{H} X_{E\_b}^h \times C_{E\_unit}^h \quad [5.3]$$

Whether the optimization concerns a building, or a building group analysis, for the evaluation of the GA-based results, a comparison to baseline consumption – as obtained by the ANN day-ahead prediction – is conducted. The cost linked to the GA-optimized solution is compared with the cost of the baseline scenario and given by the generic equations [5.4] and [5.5], respectively

$$Cost_{E\_opt} = \sum_{h=1}^{H} (X_{E_{opt}}^h \times C_{E_{unit}}^h) \quad [5.4]$$

$$Cost_{E\_baseline} = \sum_{h=1}^{H} (X_{E_{baseline}}^h \times C_{E_{unit}}^h) \quad [5.5]$$

At the building group level, the load shifting term of the objective function in equation [5.1] is calculated by equations [5.6] and [5.7] as follows:

$$Load_{Shift} = \sum_{b}^{B} Load_{Shift}^b \quad [5.6]$$

where:

$$Load_{Shift}^b = \sum_{h=1}^{H} |X_{E_b}^h - X_{E_{b_{baseline}}}^h| \quad [5.7]$$

The constraint in equation [5.8] is applied to ensure that there is no deviation between the total daily energy consumed between baseline and optimized solutions for each building:

$$\sum_{h=1}^{H} X_{E_b}^{h} - \sum_{h=1}^{H} X_{E_{b_{baseline}}}^{h} = 0 \qquad [5.8]$$

Finally, constraints on the hourly energy consumption of the optimized solution are applied to enable preferences or limitations in shifting loads from one time period within the day to another, as shown in equation [5.9].

$$X_{Emin_b}^{h} \leq X_{E_b}^{h} \leq X_{Emax_b}^{h} \qquad [5.9]$$

5.3. ToU case study

5.3.1. *ANN-based predictions*

The results of ANN-based net electrical power predictions for the period from 1/2/17 to 28/4/17 (first period), 2/5/17 to 1/8/2017 (second period) and 2/8/17 to 29/11/17 (third period) for L2 (Summa), L4 (Leaf Lab) and L5 (Kite Lab) are presented in Figures 5.3–5.5, respectively. The day-ahead predicted values correspond to a Pearson's correlation for L2 ranging from 0.91 to 0.97, L4 close to 0.4 and L5 between 0.97 and 0.98 for training, validation, testing and overall. Furthermore, the ANN-based predictions of net electrical energy consumption of the buildings under study, versus the actual measured values for a working week in the summer from 24/7/17 to 28/7/17 (top) and a working week in the winter from 20/11/17 to 24/11/17 (bottom), are illustrated in Figure 5.6. It is observed that the predicted time series largely coincides with the measured (actual) time series for both periods, and all three buildings under investigation.

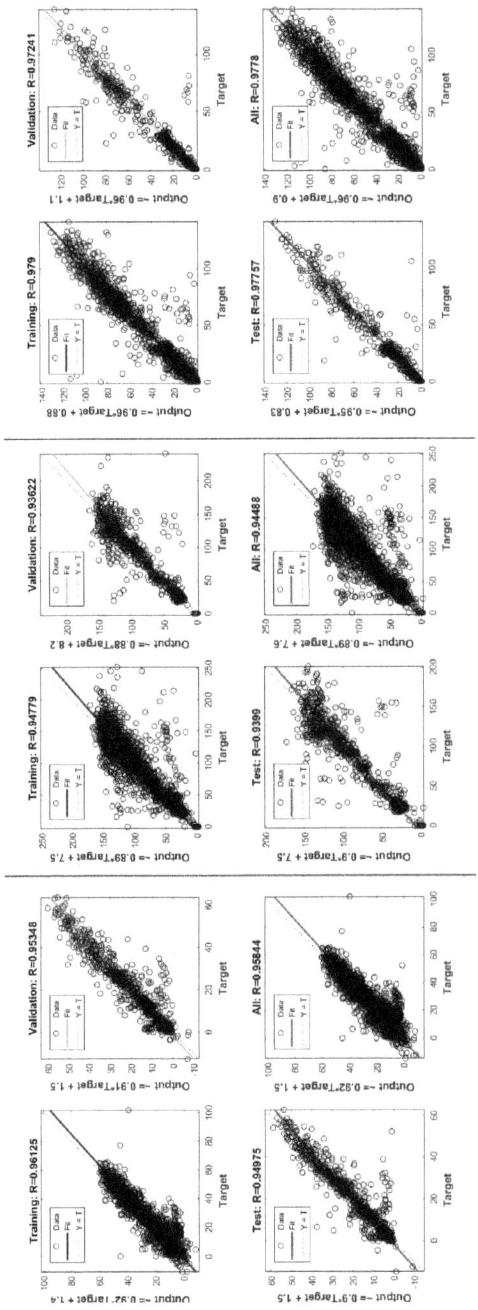

Figure 5.3. *Prediction of net electrical power consumption of L2, L4 and L5 for the first period of 2017. For a color version of this figure, see www.iste.co.uk/kolokotsa/smartbuildings.zip*

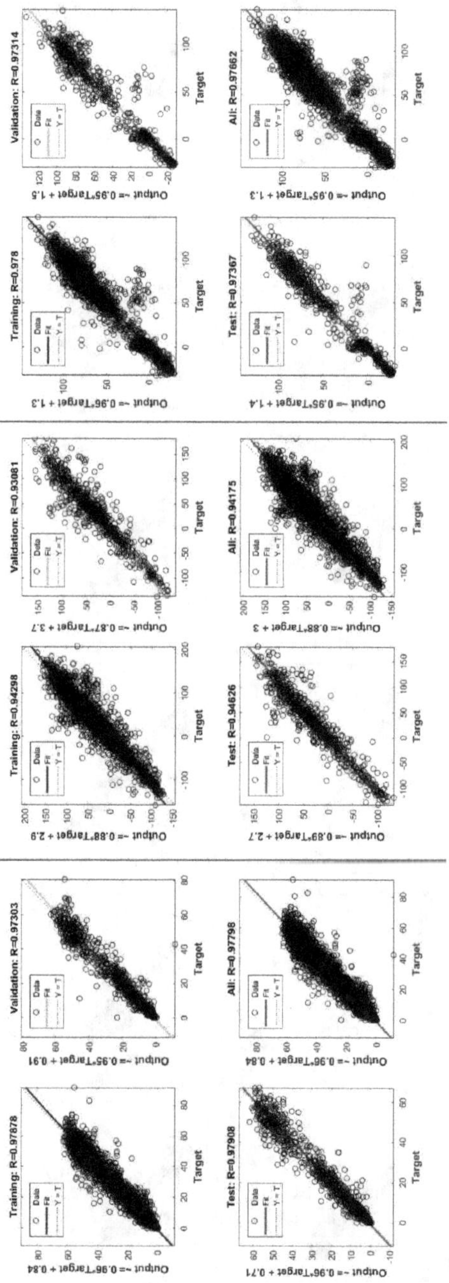

Figure 5.4. *Prediction of net electrical power consumption of L2, L4 and L5 for the second period of 2017. For a color version of this figure, see www.iste.co.uk/kolokotsa/smartbuildings.zip*

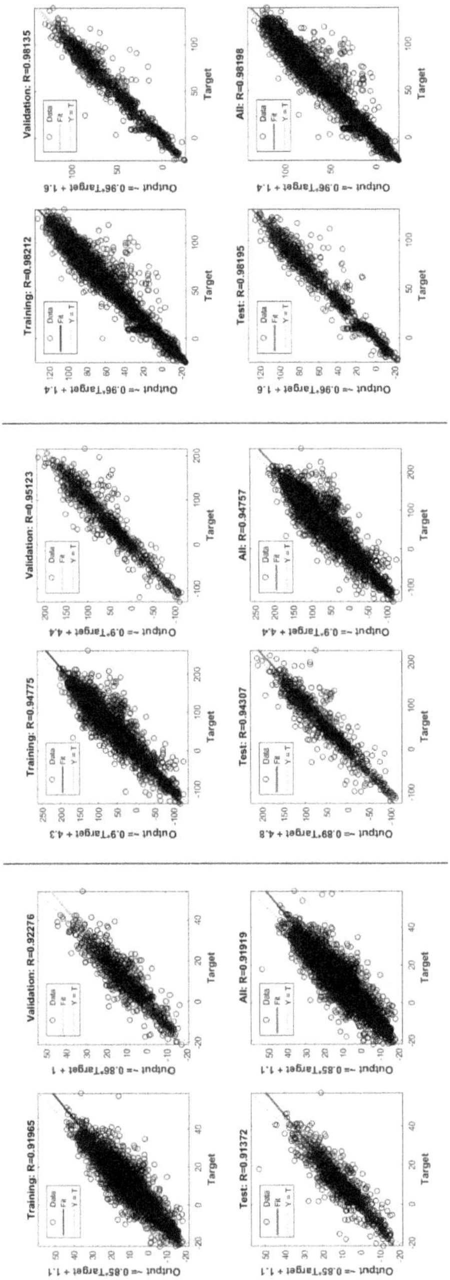

Figure 5.5. *Prediction of net electrical power consumption of L2, L4 and L5 for the third period of 2017. For a color version of this figure, see www.iste.co.uk/kolokotsa/smartbuildings.zip*

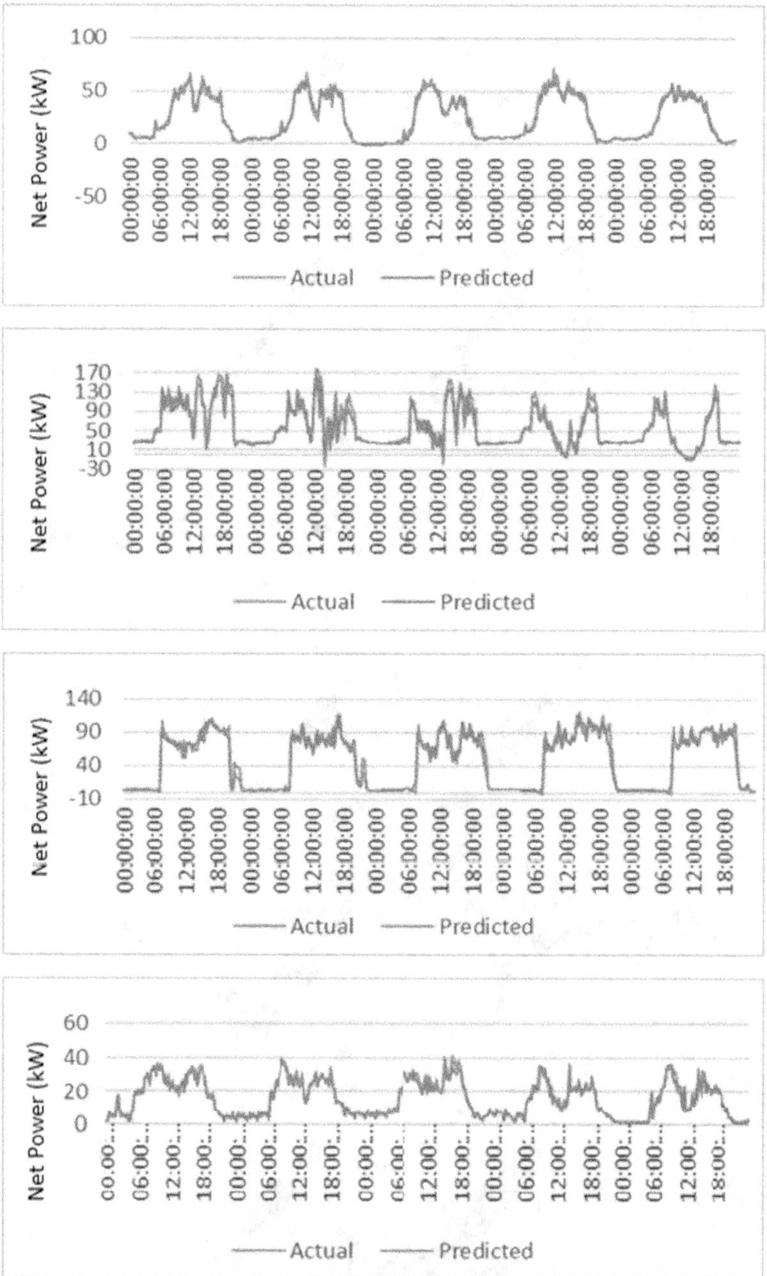
Net Power (kW)
100
50
0
-50
00:00:00
06:00:00
12:00:00
18:00:00
Actual
Predicted
Net Power (kW)
170
130
90
50
10
-30
Actual
Predicted
Net Power (kW)
140
90
40
-10
Actual
Predicted
Net Power (kW)
60
40
20
0
Actual
Predicted

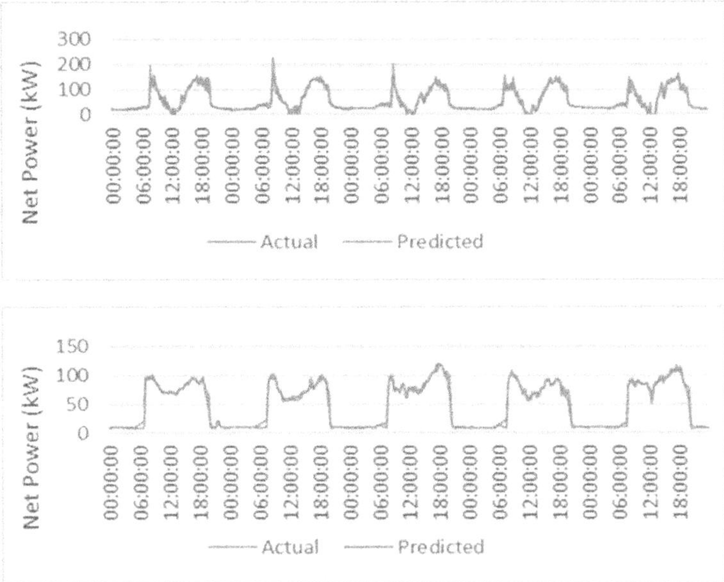

Figure 5.6. *Prediction of net electrical power consumption for L2, L4 and L5 from 24/7/17 to 28/7/17 (top) and from 20/11/17 to 24/11/17 (bottom). For a color version of this figure, see www.iste.co.uk/kolokotsa/smartbuildings.zip*

Mean bias error (MBE) and mean absolute percentage error (MAPE) values for the ANN predicted versus actual values for the periods from 24/7/17 to 28/7/17, and from 20/11/17 to 24/11/17 for Summa, Leaf Lab and Kite Lab, are presented in Table 5.2. MAPE values in Table 5.2 are notably increased by a range of ratios of actually low numerator differences, divided by denominators that approximate to zero.

ANN prediction	24/7/17 to 28/7/17		20/11/17 to 24/11/17	
	MBE	MAPE (%)	MBE	MAPE (%)
L2: Summa	0.21	32.62	–0.52	12
L4: Leaf Lab	–3	29	–0.40	20.73
L5: Kite Lab	–0.94	35.32	–0.01	11

Table 5.2. *MBE and MAPE for ANN predictions*

5.3.2. *GA optimization results*

In this section, GA optimization results for 24/7/17 and 20/11/17 are presented and analyzed for the weighting coefficient values $w_1 = w_2 = 0.5$. For the baseline scenario, a flat tariff at 0.28 €/kWh is used (Flat 1, Figure 5.7). The optimized scenario is calculated taking into account a two-zone tariff ToU pricing scheme of 0.2 €/kWh from 8am to 6pm and 0.30 €/kWh from 6pm to 8am (ToU1, Figure 5.7).

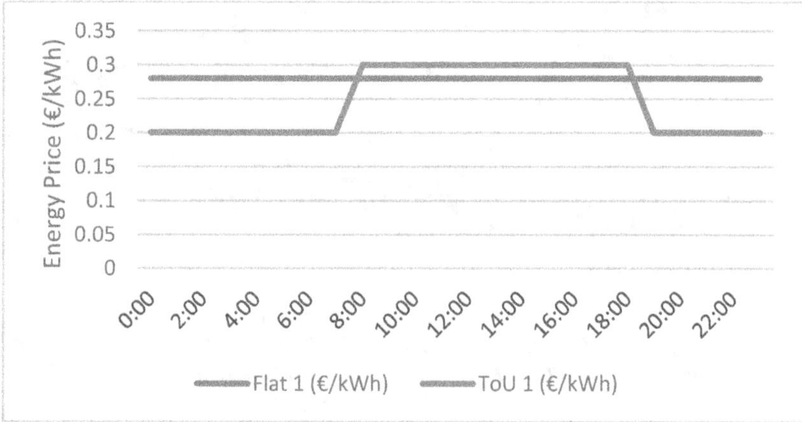

Figure 5.7. *Energy pricing profiles used in the baseline and optimized scenarios*

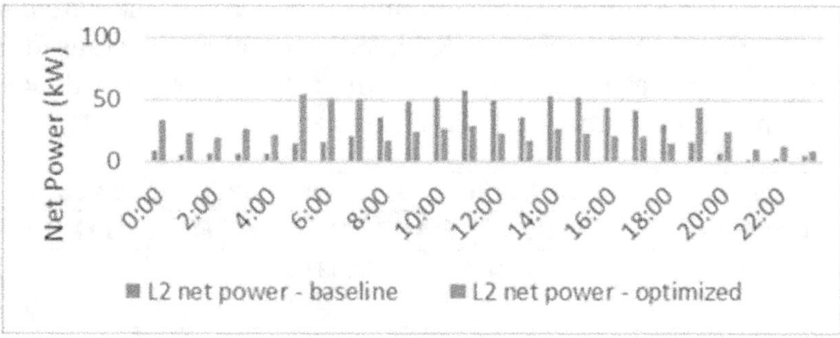

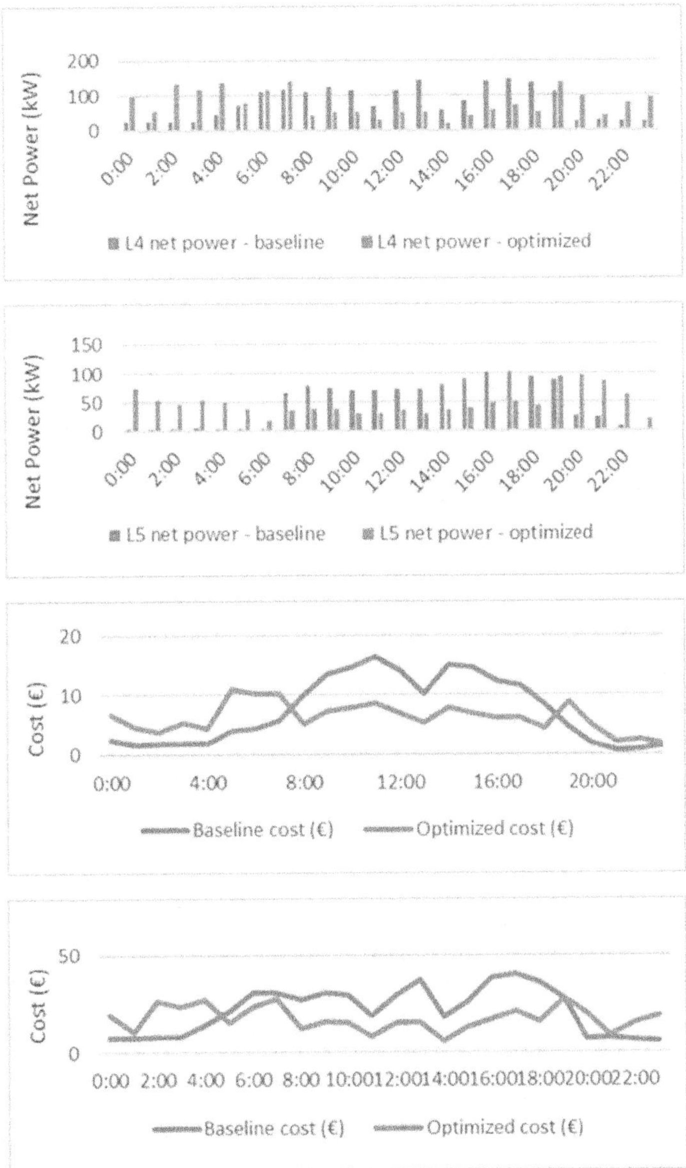
Net Power (kW)
200
100
0
0:00 2:00 4:00 6:00 8:00 10:00 12:00 14:00 16:00 18:00 20:00 22:00
L4 net power - baseline
L4 net power - optimized
Net Power (kW)
150
100
50
0
0:00 2:00 4:00 6:00 8:00 10:00 12:00 14:00 16:00 18:00 20:00 22:00
L5 net power - baseline
L5 net power - optimized
Cost (€)
20
10
0
0:00 4:00 8:00 12:00 16:00 20:00
Baseline cost (€)
Optimized cost (€)
Cost (€)
50
0
0:00 2:00 4:00 6:00 8:00 10:0012:0014:0016:0018:0020:0022:00
Baseline cost (€)
Optimized cost (€)

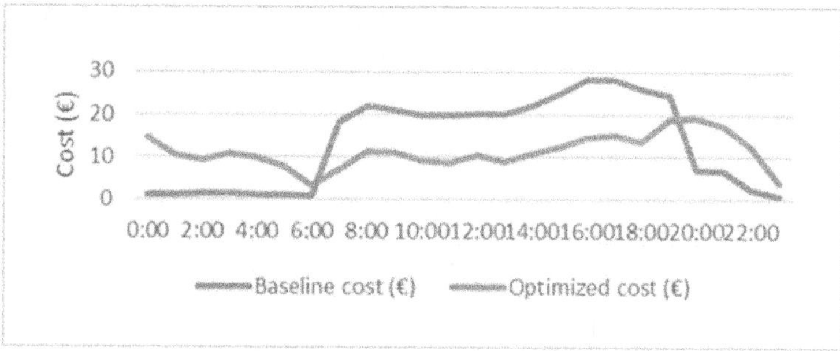

Figure 5.8. *GA optimization power and cost results for L2, L4 and L5 on 24/7/17. For a color version of this figure, see www.iste.co.uk/kolokotsa/smartbuildings.zip*

In Figure 5.8, the results of the developed GA optimization approach are presented. The top three charts illustrate the ANN-based power forecast as the baseline scenario. In these charts, the GA optimized power profiles demonstrate load shifting solutions. The related costs are depicted in the charts towards the bottom of the figure. The baseline costs are calculated based on the flat tariff shown in Figure 5.7, while the GA optimized costs are based on the two-zone tariff.

With respect to the net electrical consumption of the L2 building, it is observed in Figure 5.8 that load shifting occurs from the high-price to low-price hours. This is also reflected, in terms of the cost profile, to the day which accounts for a reduction of 15.08% from € 173.49 to €147.32. Likewise, the net electrical consumption in L4 is shifted outside the high-price region, with the baseline daily cost of €515.71 decreasing to €420.06, a reduction of 13.73%. Similarly, shifting of net electrical energy consumption in L5 occurs from the high-tariff zone toward the early morning and the evening hours, without a reduction in total energy consumption. In this case, the baseline cost is €321.29 and the optimized total cost is €271.74, which is equal to a reduction of 15.42%.

Analysis of the winter results is displayed in Figure 5.9. The shift for the L2 net electrical power profile leads to a cost reduction of 17.3%, from €123.26 in the baseline scenario down to €101.92.

Load shifting throughout the 24 hours occurs in L4 in a way that changes the overall power profile, especially with respect to the early hours of the day. This transition of loads, corresponds to 18.09% of cost savings, also reflecting the differences between the flat and the two-zone tariff pricing scheme.

With respect to daily power in L5 during the winter, changes between baseline and optimized scenarios appear to take place in a harmonic way from high- to low-price hours. In this case, a 17.55% cost saving is achieved, as the baseline daily cost is €331.53 compared with the optimized daily cost of €273.33.

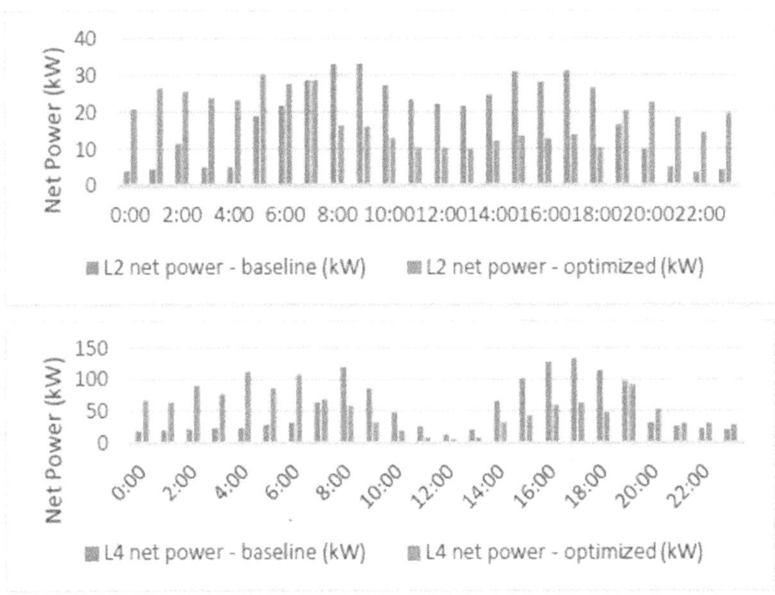

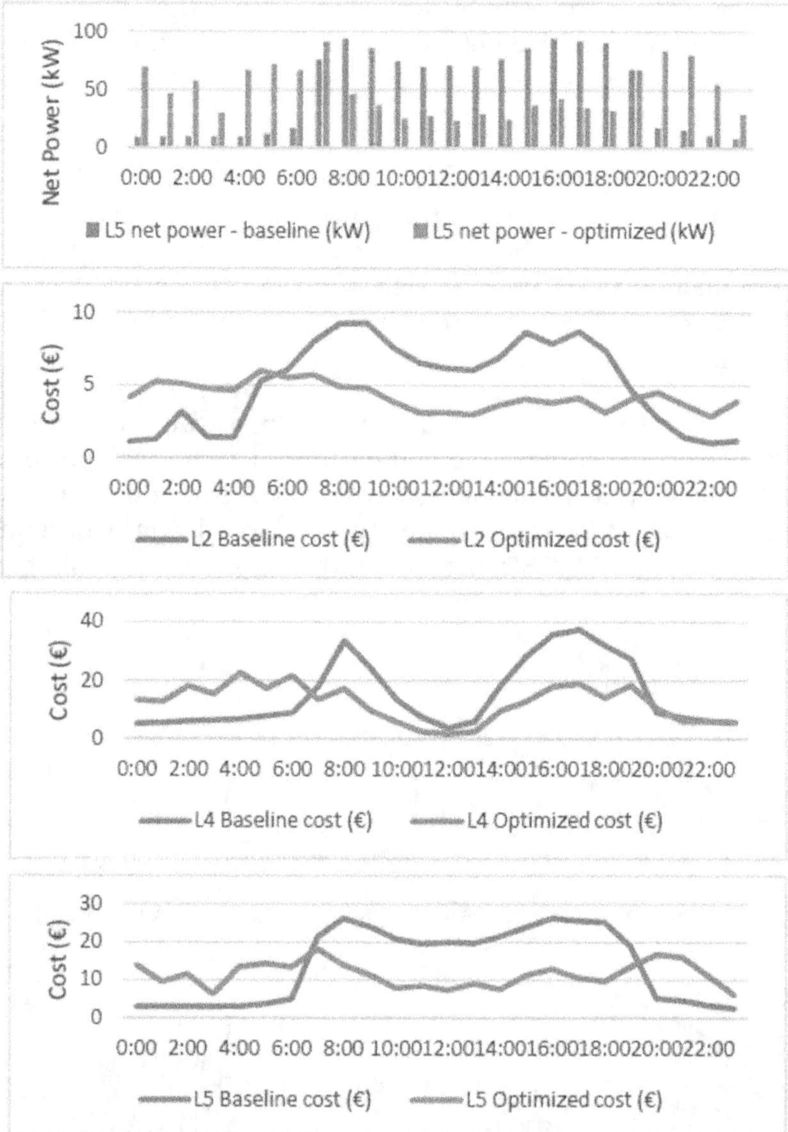

Figure 5.9. *GA optimization power and cost results for the Leaf Lab, the Summa and the Kite Lab during 20/11/17. For a color version of this figure, see www.iste.co.uk/kolokotsa/smartbuildings.zip*

In Figure 5.10, the total power consumption of the three buildings is illustrated. In the first case, the high power consumption according to the baseline power is shifted from working hours toward early morning and late evening hours. In terms of cost, the total baseline cost at the district level is €1,009.67 and the total optimized cost is €835.55, which corresponds to a reduction of 17.24%.

With respect to the winter period, the hourly district level GA-optimized power values, for equal weighting coefficients, undergo a significant differentiation with respect to the baseline. The district level total baseline cost is €814.51 and the total optimized cost is €683.05, leading to a reduction of 16.13%.

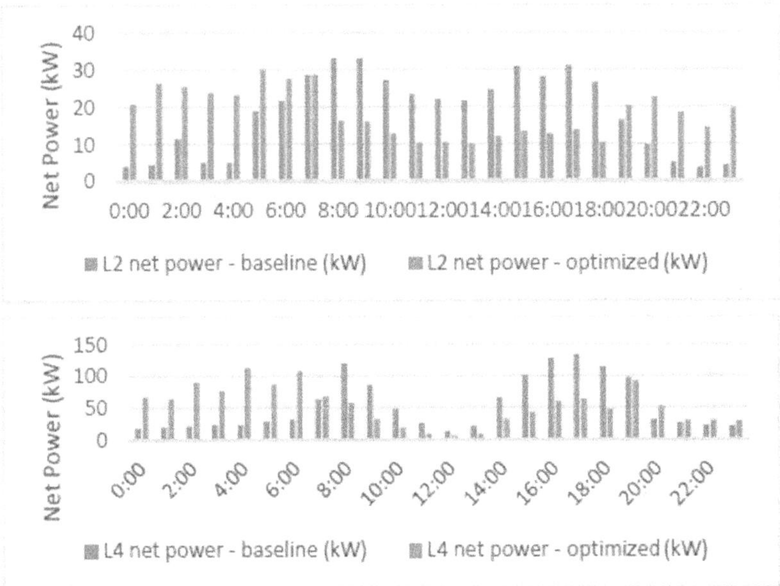

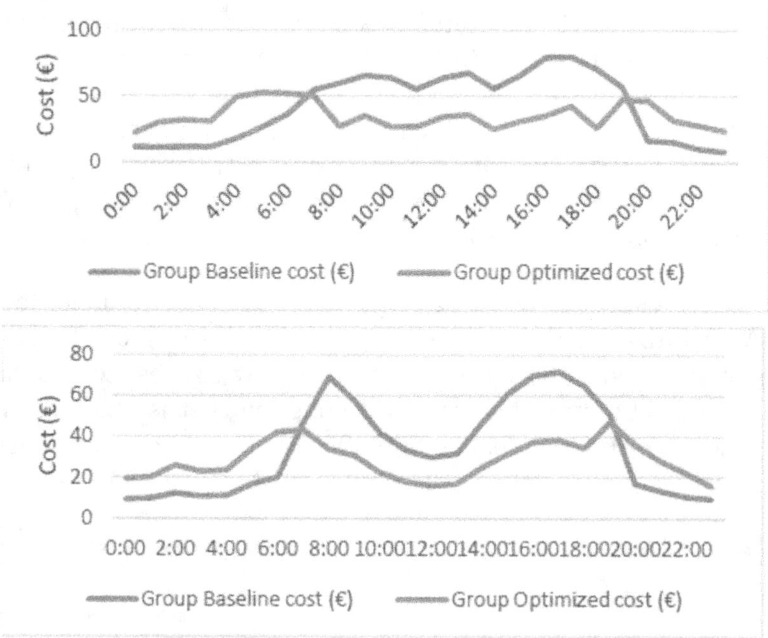

Figure 5.10. *GA optimization power and cost results for total power on 24/7/17 (up) and 20/11/17 (down). For a color version of this figure, see www.iste.co.uk/kolokotsa/smartbuildings.zip*

According to Table 5.3, in relation to the Summa building (L2), the results for each case prove that the optimization is successful, bearing in mind that the baseline cost is €172.67 and the optimized values range from €145.79 to €147.23, a maximum operational costs percentage reduction of 15.56%. For the Leaf Lab (L4), the optimized cost for each pair of weights is lower than the baseline cost of €515.71, and varies between €414.18 and €422.05. The percentage reduction, in this case, reaches 19.68%. Furthermore, the optimization for the Kite Lab revealed that the GA produces better results compared to the baseline cost of €321.29 for all pairs of weights, ranging from €269.85 down to €271.83. The percentage reduction, in this case, is up to 16.01%. The last column of the table represents the optimized cost for the group of buildings, which is lower than the baseline cost of €1,009.67 for all pairs of weighting coefficients, varying from €835.15 to €841.70. The maximum percentage reduction, in this case, is 15.39%.

w1 : Cost	w2: Load shifting	Summa (L2) cost (€)	Leaf Lab (L4) cost (€)	Kite Lab (L5) cost (€)	District level cost (€)
0	1	146.59	421.03	269.85	836.16
0.1	0.9	147.18	422.05	270.93	836.45
0.2	0.8	147.23	420.30	270.35	839.13
0.3	0.7	146.42	421.67	270.45	836.70
0.4	0.6	146.30	414.18	271.83	839.59
0.5	0.5	147.33	420.06	271.75	835.56
0.6	0.4	147.00	419.09	270.63	837.96
0.7	0.3	147.19	419.03	270.50	840.83
0.8	0.2	146.69	418.24	269.54	839.05
0.9	0.1	145.79	418.88	270.73	841.70
1	0	146.51	415.33	270.34	835.15

Table 5.3. *Results of the optimization on 24/7/17 during the summer period*

Table 5.4 includes the results of optimization for each pair of weighting coefficients at both building and district levels for the winter period. The results for the Summa building (L2) depict the optimized cost for all weight combinations. As observed, in all cases, the optimized cost varies between €100.21 and €101.92, which is lower than the baseline cost of €123.26 in this case and accounts for a percentage reduction of up to 18.70%. Moreover, the optimization for the Leaf Lab building (L4) revealed GA solutions with costs from €289.95 to €294.94, a maximum percentage reduction of 19.39% compared to the baseline cost of €359.71 in this case. Subsequently, in the Kite Lab, the optimized cost is from €277.66 down to €273.31, equal to a percentage reduction of up to 17.56% lower than the baseline cost of €331.53. The last column of the table represents the optimized cost in the group of buildings during the winter, varying from €684.77 to €682.33, leading to a maximum percentage reduction of 16.22% compared to the baseline cost of €814.51.

w_1	w_2	Summa (L2) cost (€)	Leaf Lab (L4) cost (€)	Kite Lab (L5) cost (€)	District level cost (€)
0	1	101.46	293.21	276.71	683.95
0.1	0.9	101.53	289.95	276.36	684.77
0.2	0.8	100.85	291.78	275.95	683.48
0.3	0.7	100.88	294.94	277.35	682.33
0.4	0.6	101.50	293.35	277.66	684.50
0.5	0.5	101.92	294.64	273.33	683.06
0.6	0.4	101.65	292.97	276.87	683.47
0.7	0.3	100.45	294.85	277.30	684.69
0.8	0.2	100.99	293.46	275.64	684.56
0.9	0.1	101.35	290.87	273.31	684.34
1	0	100.21	293.68	275.31	683.37

Table 5.4. *Results of the optimization on 20/11/17 during the winter period*

5.4. DA real-time case study

5.4.1. *ANN-based predictions*

ANN models are conceived on the basis of biological nervous systems to imitate information processing and evolution. ANNs assimilate the natural bonds of neurons and their high-level interconnection to model complex systems. In the case of predictions, ANNs can be more effective compared with statistical, linear or nonlinear programming techniques. ANN models have been used for years in different areas of engineering, science and business to deal with complexity and nonlinearity of data sets. They present capabilities such as adaptive learning, self-organization, real-time operation, fault tolerance and approximation of complex nonlinear functions. The mathematical model of a neuron is presented in Figure 5.11 (Shahidehpour *et al.* 2002).

Various ANN architectures for forecasting demand in electric power systems are presented by Tsekouras *et al.* (2015). A case study of the Greek electric power grid is used to showcase the performance of different ANN configurations, and factors including period length and inputs for training, confidence interval and so on. Hybrid short-term load forecasting ANN with techniques such as fuzzy logic, GA and PSO are briefly discussed in Baliyan *et al.* (2015).

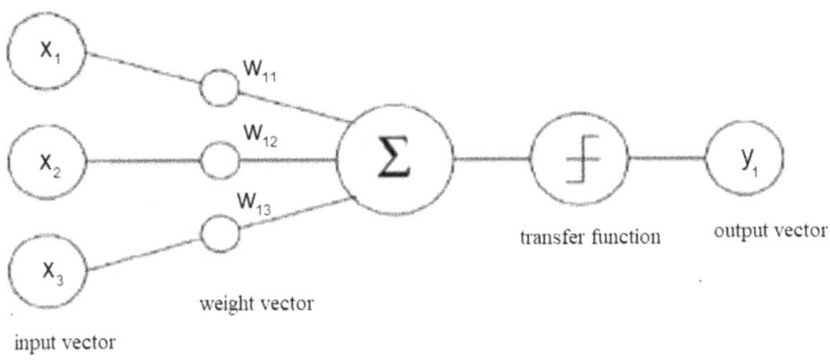

Figure 5.11. *Mathematical model of a neuron*

For the 24-hour ahead prediction of power consumption, day, time and external temperature were used as inputs with electrical power as the target. The 24-hour prediction of energy produced by renewable energy sources, day, time and irradiance were used as inputs with electrical power as the target. The Lavemberg–Marquardt algorithm was deployed in a NARX.

A summary of ANN model predictions, based on 15 minute time step data, evaluated with the aid of Pearson's correlation coefficient is provided in Table 5.5.

2/2/17-29/4/17	Pearson's coefficient R	Training	Validation	Test	Overall
15-min timestep	L2 consumption	0.96518	0.95361	0.96684	0.96376
	L2 production	0.95796	0.9325	0.94017	0.9513
	L4 consumption	0.95665	0.9553	0.95135	0.95568
	L4 production	0.98507	0.97854	0.97351	0.98236
	L5 consumption	0.98261	0.97911	0.9719	0.98058
	L5 production	0.98585	0.97547	0.98528	0.98426
	Microgrid consumption	0.98529	0.98534	0.98593	0.98539
	Microgrid production	0.98343	0.97897	0.98202	0.98254
2/5/17-1/8/17	Pearson's coefficient R	Training	Validation	Test	Overall
15-min timestep	L2 consumption	0.95152	0.95341	0.95072	0.95166
	L2 production	0.95546	0.96012	0.95837	0.95656

	L4 consumption	0.97811	0.97871	0.97204	0.97729
	L4 production	0.98059	0.98496	0.97866	0.98096
	L5 consumption	0.98184	0.97779	0.97659	0.98048
	L5 production	0.98196	0.98104	0.9806	0.98162
	Microgrid consumption	0.98982	0.99138	0.98869	0.98991
	Microgrid production	0.9815	0.98103	0.98368	0.98177
2/8/17-29/11/17	Pearson's coefficient R	Training	Validation	Test	Overall
15-min timestep	L2 consumption	0.95181	0.95267	0.95787	0.95281
	L2 production	0.95604	0.9486	0.95174	0.95427
	L4 consumption	0.97573	0.97283	0.97241	0.9748
	L4 production	0.97778	0.97099	0.97594	0.97648
	L5 consumption	0.98024	0.98066	0.98115	0.98044
	L5 production	0.9768	0.97723	0.97818	0.97707
	Microgrid consumption	0.98955	0.98862	0.98863	0.98928
	Microgrid production	0.97814	0.98149	0.98181	0.97918

Table 5.5. *Summary of ANN predictions (Pearson's correlation coefficient R) for a 15-minute timestep*

Similarly, for a timestep of 1 hour, correlation of training, validation, test and overall prediction with real values is presented in Table 5.6.

2/2/17-29/4/17	Pearson's coefficient R	Training	Validation	Test	Overall
1-hour timestep	L2 consumption	0.96129	0.95094	0.9418	0.95685
	L2 production	0.95145	0.96731	0.96268	0.95534
	L4 consumption	0.94398	0.90304	0.91239	0.9332
	L4 production	0.9696	0.95827	0.95994	0.96635
	L5 consumption	0.97321	0.95967	0.95715	0.96859
	L5 production	0.9785	0.96903	0.96748	0.97536
	Microgrid consumption	0.98456	0.97903	0.97272	0.98185
	Microgrid production	0.97633	0.96367	0.97019	0.97358
2/5/17-1/8/17	Pearson's coefficient R	Training	Validation	Test	Overall
1-hour timestep	L2 consumption	0.96568	0.95888	0.96533	0.9646
	L2 production	0.95845	0.94951	0.9684	0.95848
	L4 consumption	0.98329	0.97145	0.97276	0.97991
	L4 production	0.97867	0.97159	0.97193	0.97653
	L5 consumption	0.97935	0.97571	0.97412	0.97791
	L5 production	0.97842	0.97549	0.96029	0.97517
	Microgrid consumption	0.99136	0.98762	0.98968	0.99051
	Microgrid production	0.97754	0.97458	0.96684	0.97559
2/8/17-30/10/17	Pearson's coefficient R	Training	Validation	Test	Overall
1-hour timestep	L2 consumption	0.95021	0.94759	0.93891	0.94792
	L2 production	0.97436	0.96261	0.96707	0.97168
	L4 consumption	0.96217	0.96687	0.9598	0.96251
	L4 production	0.9754	0.96104	0.98115	0.97429
	L5 consumption	0.98388	0.98042	0.97804	0.98241

	L5 production	0.9725	0.97069	0.97084	0.97193
	Microgrid consumption	0.98987	0.98547	0.99004	0.98921
	Microgrid production	0.9771	0.97756	0.97273	0.97643
2/11/17-30/12/17	Pearson's coefficient R	Training	Validation	Test	Overall
1-hour timestep	L2 consumption	0.95817	0.95945	0.9479	0.95677
	L2 production	0.95075	0.9141	0.90309	0.93871
	L4 consumption	0.95108	0.94872	0.9343	0.94781
	L4 production	0.96894	0.9123	0.92778	0.95574
	L5 consumption	0.96995	0.95004	0.95726	0.96491
	L5 production	0.93863	0.95771	0.93775	0.94213
	Microgrid consumption	0.98859	0.98068	0.98116	0.98624
	Microgrid production	0.95557	0.94048	0.93984	0.95111

Table 5.6. *Summary of ANN predictions (Pearson's correlation coefficient R) for a timestep of 1 hour*

With respect to the quality of the prediction, we can identify differences due to various reasons. The timestep seems to be a factor that slightly affects the quality of the prediction according to R values in Tables 5.5 and 5.6. Even if not in all cases, a 15-minute timestep normally provides better prediction results compared to a 1-hour timestep. This can be attributed to a higher resolution leading to improved training of the ANN model. Another observation is that power consumption of buildings L4, L5 and the microgrid are more predictable than L2. This is possibly related to the variability and stochastic nature of loads in L2. Finally, it is observed that the period of analysis plays an important role with respect to the outcome of the prediction. For example, in Table 5.5 the prediction of consumption in L4 during the period from 2/2/17 to 29/4/17 has an overall *R* value of 0.95568, whereas the same building, in the period from 2/5/17 to 1/8/17, has an overall *R* value of 0.97729. The reason behind this difference could be the variability of loads linked to higher variation in weather conditions. In some cases, quality of data is also an issue and this is not always easy to identify in *R* values or correlation plots but may become obvious when plotting time series data.

5.4.2. *Combined ANN prediction/GA optimization results*

5.4.2.1. *DARTP scenario 1: net microgrid level prediction and optimization – 20/3/17*

In Figure 5.12, real versus predicted power for the net electrical power withdrawn by the microgrid is presented. Daily actual net energy consumption, in this case, is 2,875.21 kWh corresponding to a cost of – according to the considered DA scheme – €163.75. The equivalent predicted values are 2,824.64 kWh and €161.94, respectively. The percentage difference between the predicted and actual energy on the day, and between the cost of energy, is 1.7% and 1.1%, respectively.

In Figure 5.13, the obtained GA solution shown is associated with significant load shifting. In detail, load shifting occurs mainly in hours 5–6, 8–11 and 12–21. The daily cost of energy, in this case, is reduced from €161.75 to €152.73 and equal to a percentage cost reduction of 5.7%.

In Figure 5.14, the graphical representation of the cost of electrical energy, according to the examined scenario, is shown. It is illustrated that higher cost savings occur during the hours of high energy prices and especially from 17:00 to 20:00.

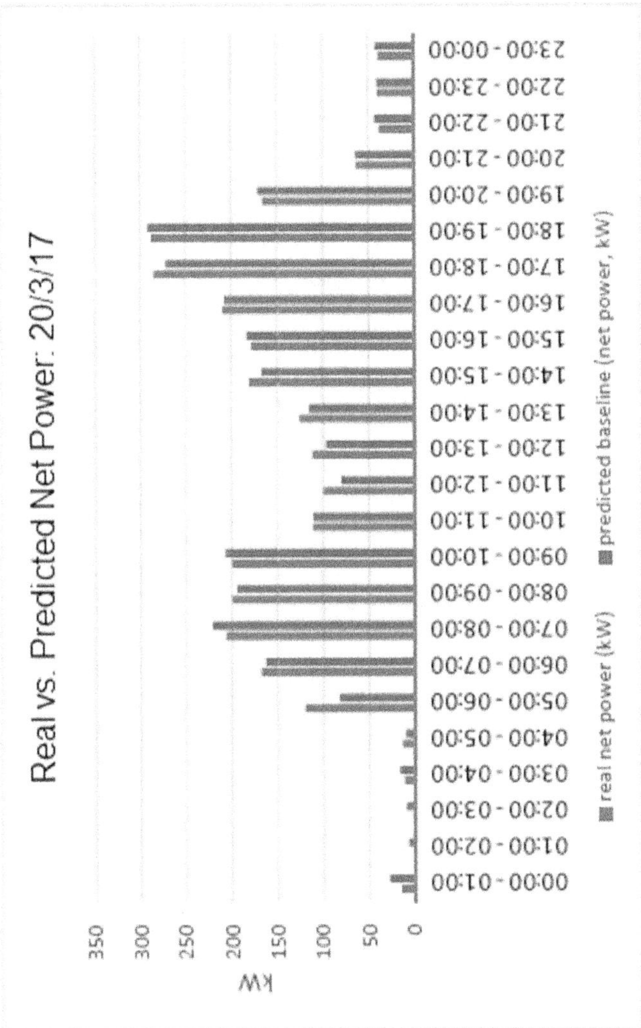

Figure 5.12. *Real versus predicted net microgrid electrical power on 20/3/17*

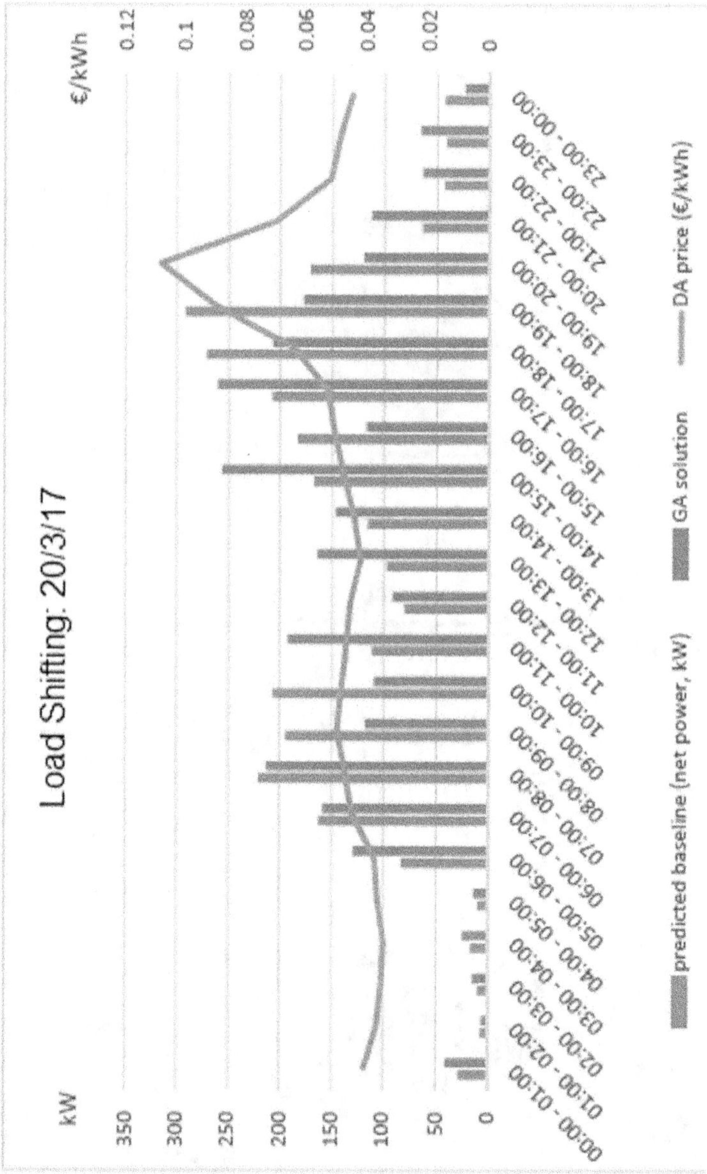

Figure 5.13. *GA obtained load shifting solution for 20/3/17*

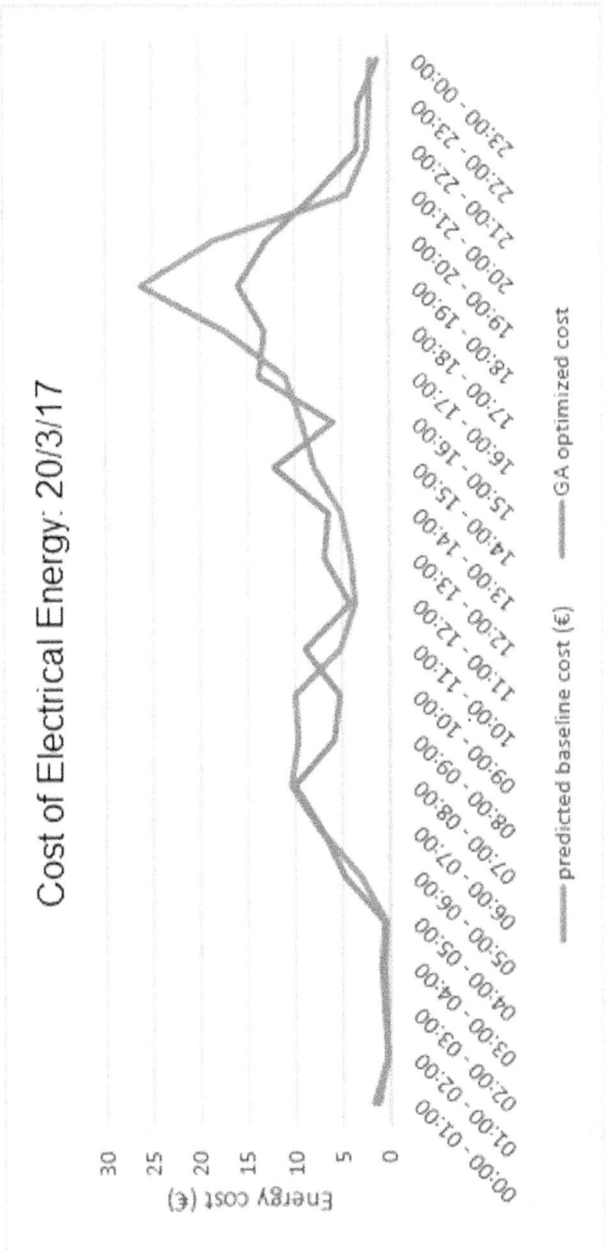

Figure 5.14. *Cost of electrical energy based on the DARTP scheme, as obtained by the GA, for 20/3/17*

5.4.2.2. *DARTP scenario 2: net microgrid level prediction and optimization – 1/8/17*

In Figure 5.15, real versus predicted power for the net electrical power withdrawn by the microgrid, for 1/8/17, is presented. Daily actual net energy consumption, in this case, is 5,586.82 kWh corresponding to a cost of – according to the considered DA scheme – €389.37. The equivalent predicted values are 5,555.08 kWh and €387.26, respectively. The percentage difference between the predicted and actual energy on the day, and between the cost of energy, is 0.57% and 0.54%, respectively.

In Figure 5.16, the obtained GA solution shown is associated with significant load shifting. In detail, load shifting occurs mainly in hours 5–8, 12–15, 14–20. The daily cost of energy, in this case, is reduced from €387.26 to €356.57, and equal to a percentage cost reduction of 7.9%.

In Figure 5.17, the graphical representation of the cost of electrical energy, according to the examined scenario, is shown. It is illustrated that higher cost savings occur during the hours of high energy prices and especially from 16:00 to 21:00.

5.4.2.3. *DARTP scenario 3a: net microgrid level prediction and optimization – 14/11/17*

In Figure 5.18, real versus predicted power for the net electrical power withdrawn by the microgrid, for 14/11/17, is presented. Daily actual net energy consumption, in this case, is 5,907.70 kWh corresponding to a cost of – according to the considered DA scheme – €537.59. The equivalent predicted values are 5,812.38 kWh and €530.16, respectively. The percentage difference between the predicted and actual energy on the day, and between the cost of energy, is 1.6% and 1.38%, respectively.

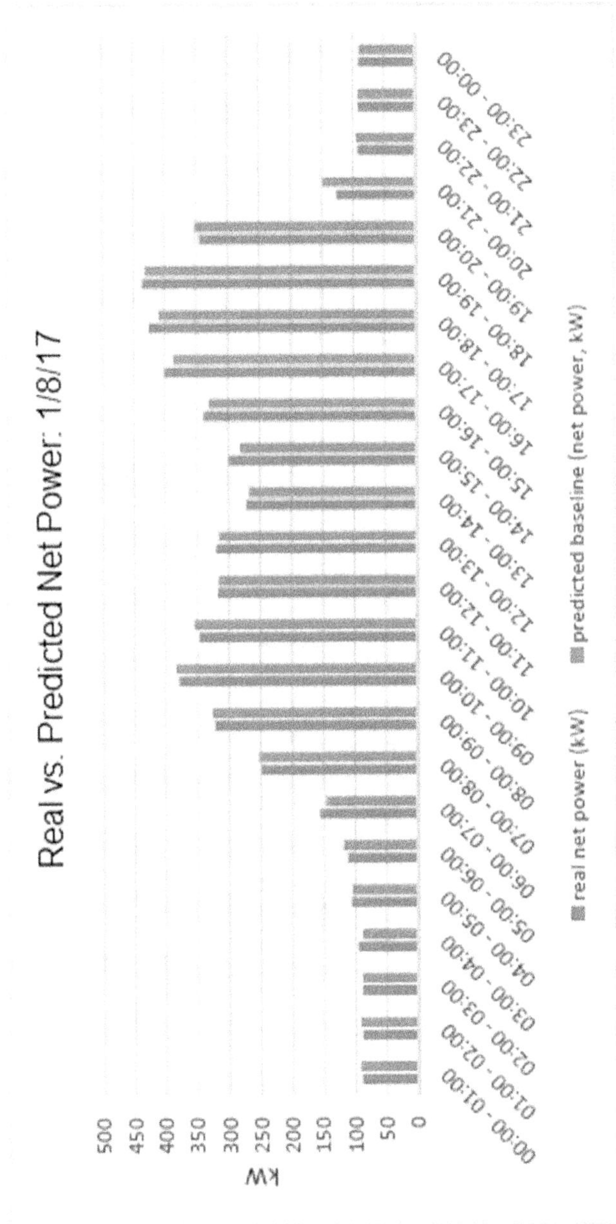

Figure 5.15. *Real versus predicted net microgrid electrical power on 1/8/17*

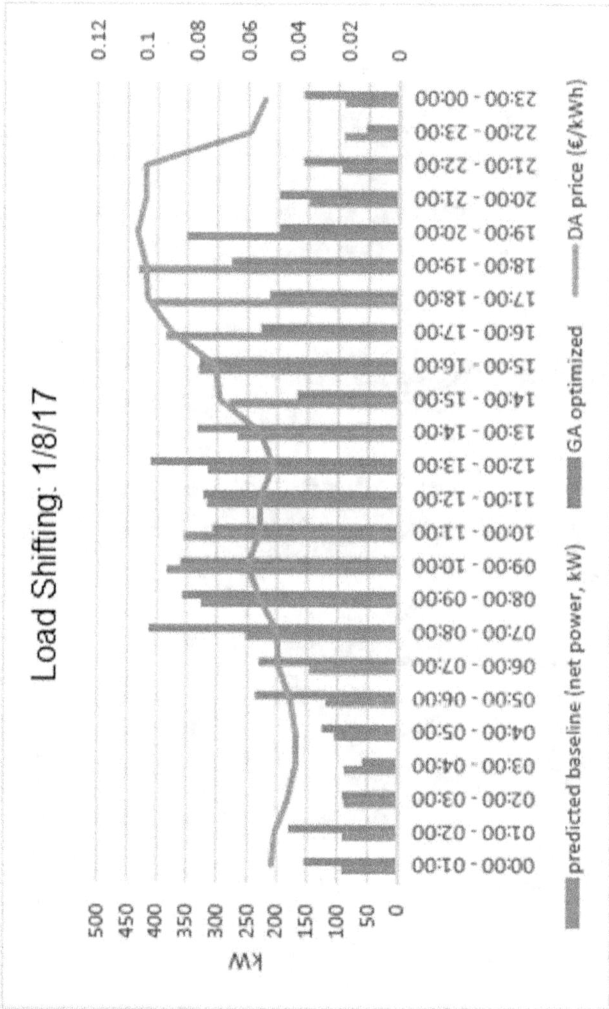

Figure 5.16. *GA obtained load shifting solution for 1/8/17. For a color version of this figure, see www.iste.co.uk/kolokotsa/smartbuildings.zip*

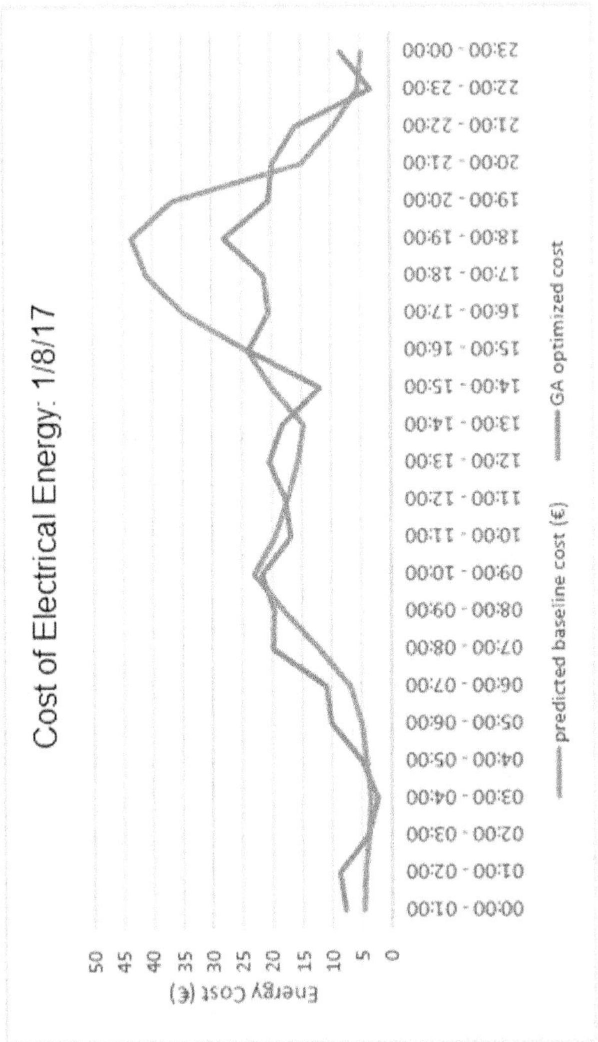

Figure 5.17. *Cost of electrical energy based on the DARTP scheme, as obtained by the GA, for 1/8/17*

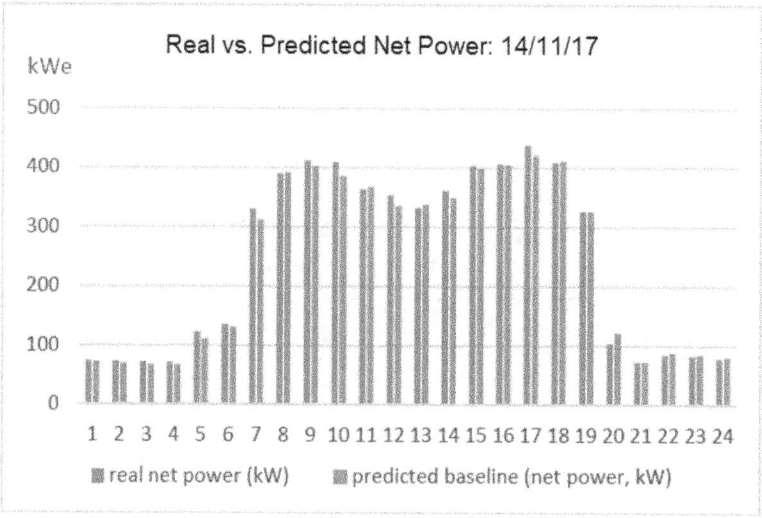

Figure 5.18. *Real versus predicted net microgrid electrical power on 14/11/17*

In Figure 5.19, the obtained GA solution shown is associated with significant load shifting. In detail, load shifting occurs mainly in hours 6–7, 9–13 and 18–21. The daily cost of energy, in this case, is reduced from €530.16 to €500.28, and equal to a percentage cost reduction of 5.6%.

In Figure 5.20, the graphical representation of the cost of electrical energy, according to the examined scenario, is shown. It is illustrated that higher cost savings occur during the hours of high energy prices and especially from 9:00 to 10:00 and 18:00 to 21:00.

5.4.2.4. *DARTP scenario 3b: net microgrid level prediction and optimization – 14/11/17*

In Figure 5.21, the obtained GA solution shown is associated with significant load shifting. In detail, load shifting occurs in hours 1, 3–5, 7, 11, 13–15, 17–19, 21 and 23–24. The daily cost of energy, in this case, is reduced from €530.16 to €502.83, and equal to a percentage cost reduction of 5.1%.

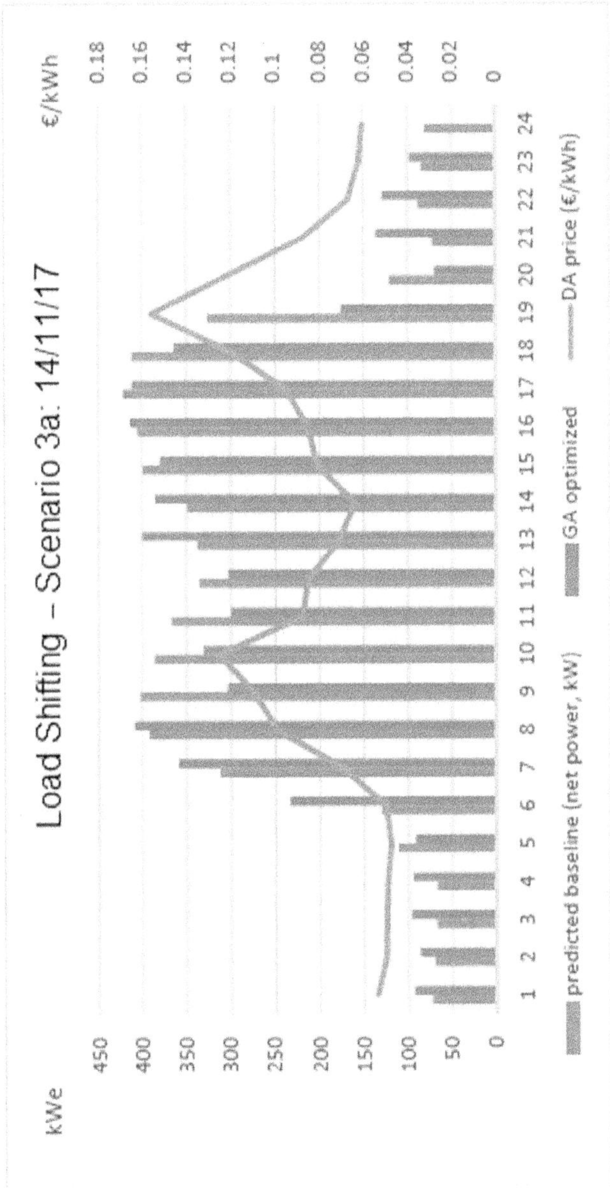

Figure 5.19. *GA obtained load shifting solution for 14/11/17. For a color version of this figure, see www.iste.co.uk/kolokotsa/smartbuildings.zip*

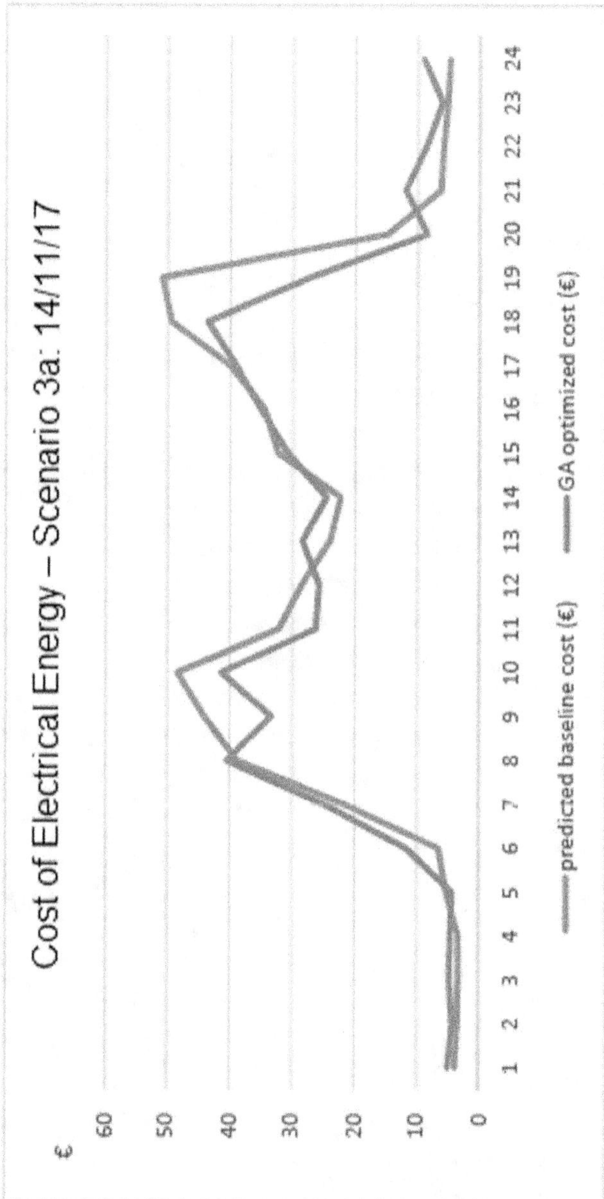

Figure 5.20. *Cost of electrical energy based on the DARTP scheme, as obtained by the GA, for 14/11/17*

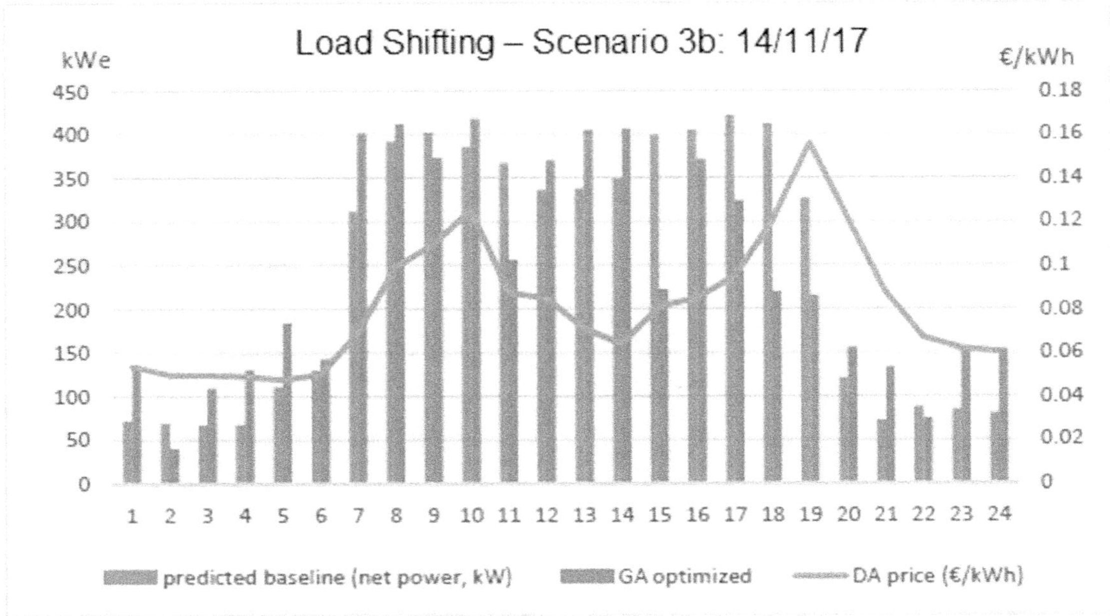

Figure 5.21. *GA obtained load shifting solution for 14/11/17. For a color version of this figure, see www.iste.co.uk/kolokotsa/smartbuildings.zip*

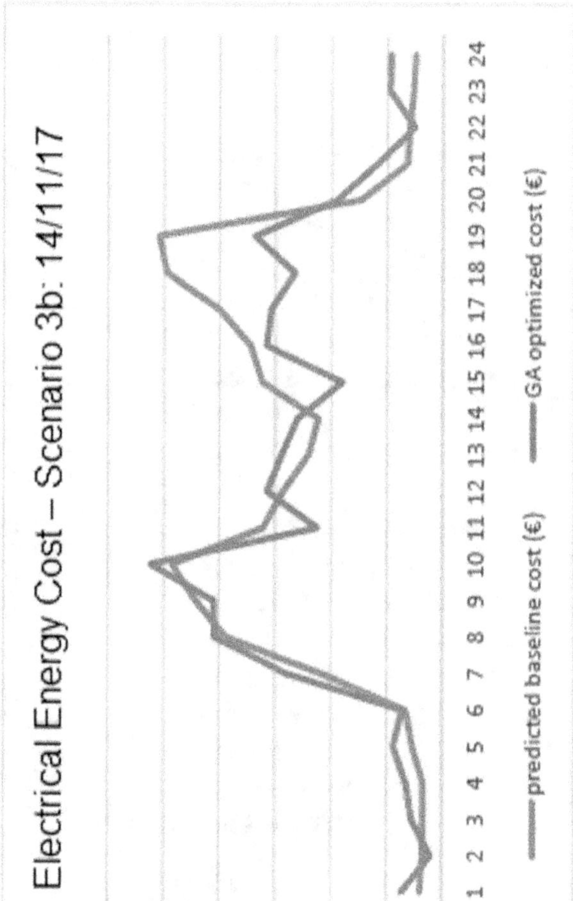

Figure 5.22. *Cost of electrical energy based on the DARTP scheme, as obtained by the GA, for 14/11/17*

In Figure 5.22, the graphical representation of the cost of electrical energy, according to the examined scenario, is shown. It is illustrated that higher cost savings occur mainly from 15:00 to 19:00.

5.5. Limitations of the proposed approach

The proposed approach entails some level of abstraction with respect to the load shift achievable within the capacity of individual systems and components. Evaluating load shift in conjunction to a pricing scheme requires deep knowledge and depends on the specificities of each case study. In this respect, load shift is determined by technical factors, i.e. the technical characteristics of installed systems and control schemes, as well as organizational factors, i.e. the potential shift of the industrial operations within each building. Detailed knowledge of the operation of each system within a building, along with data, i.e. a power consumption profile, is not available in most cases. This logic can be applied to some extent by using constraints to ensure that a specific percentage of the power at any time remains unchanged. Consequently, optimization can be conducted based on the flexible share of the power consumption for every hour.

Furthermore, the proposed approach is linked to the accuracy of the prediction which may vary according to the building under study and other factors, i.e. load type, industrial operations and season. Therefore, it is important to evaluate the risk associated with different prediction error levels, according to the examined pricing scheme.

5.6. Conclusion

The main contribution of this work relates to the linking of ANN short-term electric forecasting and GA multiobjective optimization as a tool for generating and evaluating alternative day-ahead load shifting solutions. The first step of the proposed approach is exploiting ANN modeling for the prediction of net power consumption in a period of 24 hours ahead. Predictions of net consumption power levels using the day of the week, time of day, irradiance and external temperature as inputs were obtained for each of the three buildings of the Leaf Community (Summa, Leaf Lab and Kite Lab), as well as total energy consumption for the Leaf Community microgrid. Further predictions using day of the week, time of day and irradiance were used to conduct 24-hour ahead ANN-based power generation predictions at microgrid level. The results proved that a close correlation between predicted and actual values exists during the studied summer and winter periods, as evaluated based on correlation coefficient R for the whole period, as well as MBE and MAPE-specific days used in the optimization process.

The second step was to create an optimization function to include energy cost and load shifting using appropriate variables and constraints. The objective function was minimized using a GA to obtain solutions at individual building and building group level. Results demonstrated the effectiveness of this approach in considering alternative pricing schemes and load shifting possibilities, as a way to examine cost savings. With respect to the ToU pricing scheme examined, cost savings of levels between 14.67% and 19.68% at the building level were associated with significant load shifting solutions, obtained by the GA scheme in the two-zone ToU pricing scheme considered. At the district level, cost savings in the range of 15.92% and 17.24% were obtained. With respect to the DARTP scheme, balanced load shifting solutions associated with cost savings between 5.1% and 7.9% were obtained.

Future steps in this work may involve: (1) extending research activities to focus more on renewable energy generation and storage capabilities; (2) reforming the GA-obtained solutions in order to take into consideration actual loads (base, fixed and flexible), renewable energy production and storage; and (3) exploiting the potential for improvements in power predictions using ANN models.

Conclusions and Recommendations

Targeting near-zero energy performance in buildings involves integrated design, energy efficiency measures, renewable energy, storage, advanced intelligence and systematic user engagement. In this book, the operational performances of a residential and an industrial near-zero energy building (NZEB) have been investigated, analyzed and optimized with the use of measurements and dynamic energy modeling. The role of renewable energy systems, storage, smart monitoring and controls for the energy performance of NZEB and microgrid integration in smart grids has been qualitatively and quantitatively assessed. Specifically, renewables and storage in buildings and microgrids are highlighted as being of major importance to minimize energy demand and allow flexibility as a valuable resource asset. Smart monitoring and indoor conditions measurements have been deeply exploited to evaluate energy efficiency aspects and to enable validation of the dynamic building energy models.

Subsequently, advanced and robust building energy models are used as the basis for real-time energy management solutions to be designed, implemented and tested. In this framework, an optimization assessment framework for

---

Chapter written by Nikos KAMPELIS.

heating, ventilation and air conditioning (HVAC) energy management in day-ahead real-time pricing demand response programs was developed. Results demonstrate a strong potential for energy and cost savings based on the provided optimized control of indoor conditions while indoor thermal comfort remains within prescribed levels. The scenarios examined are associated with potential levels of cost reductions in the order between 9.9% and 25% and HVAC energy reduction between 10.4% and 25%. The selected solutions fully comply with indoor comfort and indoor temperature drift rate standards. The developed approach can be widely used due to the fact that it deploys temperature set points for HVAC energy efficiency assessment and control. It allows expandability in establishing optimal control of thermal zones in buildings of various uses and sizes controlled by single or distributed thermostatic controls. A major conclusion arising from this work is that HVAC dynamic control associated with demand response real-time pricing (RTP) schemes has high potential if intelligently integrated and explored along with the operation of smart buildings and smart grids in the near future.

Artificial neural network (ANN) short-term electric forecasting and GA multiobjective optimization have been combined to create a tool for generating and evaluating alternative day-ahead load shifting solutions. Exploiting ANN modeling has been effective for the prediction of power consumption and production in a period of 24 h ahead. Predicting hourly consumption, production and net consumption levels using appropriate input configurations has proven to be effective at the building and microgrid levels. The results proved that a close correlation between predicted and actual values exists, as evaluated based on correlation coefficient R for the whole period, as well as mean bias error (MBE) and mean average predicted error (MAPE) for specific days evaluated prior to the load shifting optimization process.

Furthermore, a GA optimization model was created to evaluate energy cost and load shifting of the ANN-predicted consumption for several scenarios at the building and microgrid levels. Power consumption and production predictions based on ANN models and GA optimization models were tested and proven to be a robust technique for the implementation of load shifting strategies and for the evaluation of energy and cost savings. Results were used to provide thorough considerations regarding the effectiveness and limitations of this approach when considering alternative pricing schemes and load shifting possibilities in order to obtain cost savings. Cost savings between 14.67% and 19.68% and in the range of 15.92% and 17.24% were associated with significant load shifting solutions for the building and district levels, respectively, when a specific two-zone time of use (ToU) scheme was considered. With respect to the day-ahead real-time pricing (DARTP) scheme, cost savings between 5.1% and 7.9% were linked to relatively balanced net microgrid level optimized solutions.

Overall, the energy and cost optimization of the operational phase of buildings demands deep knowledge of components and performance over time, coupled with intelligent advanced energy management systems. Throughout this research, a significant space of improvement in energy management both in terms of exploiting advanced control algorithms and demand response actions has been identified and demonstrated.

References

Aalami, H.A., Moghaddam, M.P., Yousefi, G.R. (2010). Demand response modeling considering interruptible/curtailable loads and capacity market programs. *Applied Energy*, 87(1), 243–250.

Abuelnasr, M., El-Khattam, W., Helal, I. (2018). Examining the influence of micro-grids topologies on optimal energy management systems decisions using genetic algorithm. *Ain Shams Engineering Journal*, 9(4), 2807–2814.

Afram, A., Janabi-Sharifi, F. (2014). Review of modeling methods for HVAC systems. *Applied Thermal Engineering*, 67(1–2), 507–519.

Afroz, Z., Shafiullah, G.M., Urmee, T., Higgins, G. (2018). Modeling techniques used in building HVAC control systems: A review. *Renewable and Sustainable Energy Reviews*, 83, 64–84.

Albadi, M.H., El-Saadany, E.F. (2008). A summary of demand response in electricity markets. *Electric Power Systems Research*, 78(11), 1989–1996.

Alhaider, M., Fan, L. (2015). Mixed integer programming for HVACs operation. *IEEE Power & Energy Society General Meeting*, Denver, 1–5.

Alharbi, T., Bhattacharya, K. (2018). A stochastic energy management system for isolated microgrids. *IEEE Power & Energy Society General Meeting*, Portland, 1–5.

Anjo, J., Neves, D., Silva, C., Shivakumar, A., Howells, M. (2018). Modeling the long-term impact of demand response in energy planning: The Portuguese electric system case study. *Energy*, 165, 456–468.

Attia, M., Haidar, N., Senouci, S.M., Aglzim, E.H. (2018). Towards an efficient energy management to reduce CO_2 emissions and billing cost in smart buildings. *15th IEEE Annual Consumer Communications & Networking Conference*, Las Vegas, 1–6.

Bahmani, R., Karimi, H., Jadid, S. (2020). Stochastic electricity market model in networked microgrids considering demand response programs and renewable energy sources. *International Journal of Electrical Power & Energy Systems,* 117.

Baliyan, A., Gaurav, K., Kumar Mishra, S. (2015). A review of short term load forecasting using artificial neural network models. *Procedia Computer Science*, 48(C), 121–125.

Batić, M., Tomašević, N., Beccuti, G., Demiray, T., Vraneš, S. (2016). Combined energy hub optimisation and demand side management for buildings. *Energy and Buildings*, 127, 229–241.

Bertoldi, P., Zancanella, P., Boza-Kiss, B. (2016). Demand response status in EU member states. JRC science for policy report, European Commission.

Brackney, L., Parker, A., Macumber, D., Benne, K. (2018). *Building Energy Modeling with OpenStudio*. Springer International Publishing, New York.

Brusco, G., Burgio, A., Menniti, D., Pinnarelli, A., Sorrentino, N., Scarcello, L. (2017). An energy box in a cloud-based architecture for autonomous demand response of prosumers and prosumages. *Electronics*, 6(4), 98.

Bui, V.-H., Hussain, A., Kim, H.-M. (2017). Optimal operation of microgrids considering auto-configuration function using multiagent system. *Energies*, 10(10), 1484.

Bui, V.-H., Hussain, A., Kim, H.-M. (2018). A multiagent-based hierarchical energy management strategy for multi-microgrids considering adjustable power and demand response. *IEEE Transactions on Smart Grid*, 9(2), 1323–1333.

Bullich-Massagué, E., Díaz-González, F., Aragüés-Peñalba, M., Girbau-Llistuella, F., Olivella-Rosell, P., Sumper, A. (2018). Microgrid clustering architectures. *Applied Energy*, 212, 340–361.

Carrasqueira, P., Alves, M.J., Antunes, C.H. (2017). Bi-level particle swarm optimization and evolutionary algorithm approaches for residential demand response with different user profiles. *Information Sciences*, 418–419, 405–420.

Cheng, L., Wan, Y., Tian, L., Tian, X. (2018). Integrated demand response under the background of energy Internet: A review and outlook. *Proceedings of the 2nd IEEE Conference on Energy Internet Energy System Integration*, Beijing, 1–6.

Comodi, G., Giantomassi, A., Severini, M., Squartini, S., Ferracuti, F., Fonti, A., Nardi Cesarini, D., Morodo, M., Polonara, F. (2015). Multi-apartment residential microgrid with electrical and thermal storage devices: Experimental analysis and simulation of energy management strategies. *Applied Energy*, 137, 854–866.

Crosbie, T., Short, M., Dawood, M., Charlesworth, R. (2017). Demand response in blocks of buildings: Opportunities and requirements. *Entrepreneurship and Sustainability Issues*, 4(3), 271–281.

Crosbie, T., Broderick, J., Short, M., Charlesworth, R., Dawood, M. (2018). Demand response technology readiness levels for energy management in blocks of buildings. *Buildings*, 8(2), 13.

D'agostino, D., Zangheri, P., Cuniberti, B., Paci, D., Bertoldi, P. (2016). Synbook report on the national plans for nearly zero energy buildings (NZEBs). Report, Publications Office of the European Union, Luxembourg.

Dawoud, S.M., Lin, X., Okba, M.I. (2018). Hybrid renewable microgrid optimization techniques: A review. *Renewable and Sustainable Energy Reviews*, 82, 2039–2052.

Deng, X., Lv, T. (2019). Power system planning with increasing variable renewable energy: A review of optimization models. *Journal of Cleaner Production*, 246, 118962.

De Rosa, M., Carragher, M., Finn, D.P. (2018). Flexibility assessment of a combined heat-power system (CHP) with energy storage under real-time energy price market framework. *Thermal Science and Engineering Progress. Prog.*, 8, 426–438.

De Wilde, P. (2014). The gap between predicted and measured energy performance of buildings: A framework for investigation. *Automation in Construction*, 41, 40–49.

Diakaki, C., Grigoroudis, E., Kolokotsa, D. (2008). Towards a multi-objective optimization approach for improving energy efficiency in buildings. *Energy and Buildings*, 40(9), 1747–1754.

Diakaki, C., Grigoroudis, E., Kabelis, N., Kolokotsa, D., Kalaitzakis, K., Stavrakakis, G. (2010). A multi-objective decision model for the improvement of energy efficiency in buildings. *Energy*, 35(12), 5483–5496.

Doostizadeh, M., Ghasemi, H. (2012). A day-ahead electricity pricing model based on smart metering and demand-side management. *Energy*, 46(1), 221–230.

Dounis, A.I., Caraiscos, C. (2009). Advanced control systems engineering for energy and comfort management in a building environment – A review. *Renewable and Sustainable Energy Reviews*, 13(6–7), 1246–1261.

EEAP (2014). Italian Energy Efficiency Action Plan.

European Commission (2011). A roadmap for moving to a competitive low carbon economy in 2050. Paper, European Commission, Brussels.

European Parliament and the Council of the European Union (2010). Directive 2010/31/EU of the European Parliament and of the Council of 19 May 2010 on the energy performance of buildings (recast). *Official Journal of the European Union*, 153, 13–35.

European Parliament and the Council of the European Union (2012). Directive 2012/27/EU of the European Parliament and of the Council of 25 October 2012 on energy efficiency, amending Directives 2009/125/EC and 2010/30/EU and repealing Directives 2004/8/EC and 2006/32/EC Text with EEA relevance. *Official Journal of the European Union,* 315, 1–56.

Fanti, M.P., Mangini, A.M., Roccotelli, M., Ukovich, W., Pizzuti, S. (2015). A control strategy for district energy management. *2015 IEEE International Conference on Automation Science and Engineering*, Gothenburg, 432–437.

Faruqui, A., Hledik, R., George, S.S., Bode, J., Mangasarian, P., Rohmund, I., Wikler, G., Ghosh, D., Yoshida, S. (2009). A national assessment of demand response potential. Report, Federal Energy Regulatory Commission, Washington, DC.

Ferrari, L., Esposito, F., Becciani, M., Ferrara, G., Magnani, S., Andreini, M., Bellissima, A., Cantù, M., Petretto, G., Pentolini, M. (2017). Development of an optimization algorithm for the energy management of an industrial Smart User. *Applied Energy*, 208, 1468–1486.

Gao, D., Sun, Y. (2016). A GA-based coordinated demand response control for building group level peak demand limiting with benefits to grid power balance. *Energy and Buildings*, 110, 31–40.

Gao, D., Sun, Y., Lu, Y. (2015). A robust demand response control of commercial buildings for smart grid under load prediction uncertainty. *Energy*, 93, 275–283.

Ghazvini, M.A.F., Soares, J., Morais, H., Castro, R., Vale, Z. (2017). Dynamic pricing for demand response considering market price uncertainty. *Energies*, 10(9), 1–20.

Gobakis, K., Mavrigiannaki, A., Kalaitzakis, K., Kolokotsa, D.-D. (2017). Design and development of a Web based GIS platform for zero energy settlements monitoring. *Energy Procedia*, 134, 48–60.

Goldman, C., Reid, M., Levy, R., Silverstein, A. (2010). Coordination of energy efficiency and demand response. Paper, National Action Plan for Energy Efficiency.

Gong, X., De Paola, A., Angeli, D., Strbac, G. (2019). A game-theoretic approach for price-based coordination of flexible devices operating in integrated energy-reserve markets. *Energy*, 189, 116153.

Good, N., Ellis, K.A., Mancarella, P. (2017). Review and classification of barriers and enablers of demand response in the smart grid. *Renewable and Sustainable Energy Reviews*, 72, 57–72.

Goulden, M., Bedwell, B., Rennick-Egglestone, S., Rodden, T., Spence, A. (2014). Smart grids, smart users? The role of the user in demand side management. *Energy Research & Social Science*, 2, 21–29.

Greer, C., Wollman, D.A., Prochaska, D.E., Boynton, P.A., Mazer, J.A., Nguyen, C.T., FitzPatrick, G.J., Nelson, T.L., Koepke, G.H., Hefner Jr., A.R., Pillitteri, V.Y., Brewer, T.L., Golmie, N.T., Su, D.H., Eustis, A.C., Holmberg, D.G., Bushby, S.T. (2014). *NIST Framework and Roadmap for Smart Grid Interoperability Standards, Release 3.0*. NIST, Gaithersburg.

Groezinger Jan, S.M., Thomas, B., John, A., Jan, S., Felix, W. (2014). Overview of member states information on NZEBs. Working version of the progress report – Final report, Ecofys, Cologne.

Guang, F., He, Y., Wen, L. (2019). Impacts of hybrid time-varying tariffs on residential electricity demand: The case of Zhejiang province. *Utilities Policy*, 61, 100969.

Guo, S., Liu, Q., Sun, J., Jin, H. (2018). A review on the utilization of hybrid renewable energy. *Renewable and Sustainable Energy Reviews*, 91, 1121–1147.

Haider, H.T., See, O.H., Elmenreich, W. (2016). Residential demand response scheme based on adaptive consumption level pricing. *Energy*, 113, 301–308.

Harish, V.S.K.V., Kumar, A. (2016). A review on modeling and simulation of building energy systems. *Renewable and Sustainable Energy Reviews*, 56, 1272–1292.

Hatziargyriou, N., Asano, H., Iravani, R., Marnay, C. (2007). Microgrids. *IEEE Power & Energy Magazine*, 5(4), 78–94.

Heinrich, C., Ziras, C., Syrri, A.L.A., Bindner, H.W. (2020). EcoGrid 2.0: A large-scale field trial of a local flexibility market. *Applied Energy*, 261, 114399.

Hemerlink Andreas, M.E., Sven, S., Thomas, B., Lorenzo, P., Paolo, Z., Roberto, A., Karsten, V. (2013). Towards nearly zero-energy buildings: Definition of common principles under the EPBD. Final report – Executive Summary, Ecofys, Cologne.

Hirsch, A., Parag, Y., Guerrero, J. (2018). Microgrids: A review of technologies, key drivers, and outstanding issues. *Renewable and Sustainable Energy Reviews*, 90, 402–411.

Huang, S., Wu, Q., Cheng, L., Liu, Z. (2016). Optimal reconfiguration-based dynamic tariff for congestion management and line loss reduction in distribution networks. *IEEE Transactions on Smart Grid*, 7(3), 1295–1303.

International Organization for Standardization (2006). Ergonomics of the thermal environment: Analytical determination and interpretation of thermal comfort using calculation of the PMV and PPD indices and local thermal comfort criteria. Standard ISO 7730:2005, ISO.

Jain, A., Mani, A., Siddiqui, A.S. (2019). Network architecture for demand response implementation in smart grid. *International Journal of Systems Assurance Engineering and Management*, 10(6), 1389–1402.

Jiang, P., Dong, J., Huang, H. (2020). Optimal integrated demand response scheduling in regional integrated energy system with concentrating solar power. *Applied Thermal Engineering*, 166, 114754.

Kalaitzakis, K., Stavrakakis, G.S., Anagnostakis, E.M. (2002). Short-term load forecasting based on artificial neural networks parallel implementation. *Electric Power Systems Research*, 63(3), 185–196.

Kalkan, N., Young, E.A., Celiktas, A. (2012). Solar thermal air conditioning technology reducing the footprint of solar thermal air conditioning. *Renewable and Sustainable Energy Reviews*, 16(8), 6352–6383.

Kampelis, N., Gobakis, K., Vagias, V., Kolokotsa, D., Standardi, L., Isidori, D., Cristalli, C., Montagnino, F.M., Paredes, F., Muratore, P., Venezia, L., Dracou, M.K., Montenon, A., Pyrgou, A., Karlessi, T., Santamouris, M. (2017). Evaluation of the performance gap in industrial, residential and tertiary near-zero energy buildings. *Energy and Buildings*, 148, 58–73.

Karlessi, T., Kampelis, N., Kolokotsa, D., Assimakopoulos, M., Santamouris, M. (2016). Towards sustainable and smart communities: Integrating energy efficient technologies into buildings through a holistic approach. *9th International Conference Improving Energy Efficiency in Commercial Buildings and Smart Communities*, Frankfurt, 920–927.

Karlessi, T., Kampelis, N., Kolokotsa, D., Santamouris, M., Standardi, L., Isidori, D., Cristalli, C. (2017). The concept of smart and NZEB buildings and the integrated design approach. *Procedia Engineering*, 180, 1316–1325.

Khan, A.R., Mahmood, A., Safdar, A., Khan, Z.A., Khan, N.A. (2016). Load forecasting, dynamic pricing and DSM in smart grid: A review. *Renewable and Sustainable Energy Reviews*, 54, 1311–1322.

Koliou, E. (2016). Demand response policies for the implementation of smart grids. PhD Thesis, Delft University of Technology.

Kolokotsa, D. (2016). The role of smart grids in the building sector. *Energy and Buildings*, 116, 703–708.

Kolokotsa, D., Diakaki, C., Grigoroudis, E., Stavrakakis, G., Kalaitzakis, K. (2009a). Decision support methodologies on the energy efficiency and energy management in buildings. *Advances in Building Energy Research*, 3(1), 121–146.

Kolokotsa, D., Pouliezos, A., Stavrakakis, G., Lazos, C. (2009b). Predictive control techniques for energy and indoor environmental quality management in buildings. *Building and Environment*, 44(9), 1850–1863.

Kolokotsa, D., Rovas, D., Kosmatopoulos, E., Kalaitzakis, K. (2011). A roadmap towards intelligent net zero- and positive-energy buildings. *Solar Energy*, 85(12), 3067–3084.

Kolokotsa, D., Gobakis, K., Papantoniou, S., Georgatou, C., Kampelis, N., Kalaitzakis, K., Vasilakopoulou, K., Santamouris, M. (2016). Development of a web based energy management system for university campuses: The CAMP-IT platform. *Energy and Buildings*, 123, 119–135.

Kolokotsa, D., Kampelis, N., Mavrigiannaki, A. (2019). On the integration of the energy storage in smart grids: Technologies and applications. *Energy Storage*, 1–25.

Koutroulis, E., Kalaitzakis, K. (2006). Design of a maximum power tracking system for wind-energy-conversion applications. *IEEE Transactions on Industrial Electronics*, 53(2), 486–494.

Koutroulis, E., Kolokotsa, D., Potirakis, A., Kalaitzakis, K. (2006). Methodology for optimal sizing of standalone photovoltaic/wind generator systems using genetic algorithm. *Solar Energy*, 80(9), 1072–1088.

Koutroulis, E., Kolokotsa, D., Stravrakakis, G. (2009). Optimal design and economic evaluation of a battery energy storage system for the maximization of the energy generated by wind farms in isolated electric grids, *Wind Engineering*, 33(1), 55–81.

Kusiak, A., Tang, F., Xu, G. (2011). Multi-objective optimization of HVAC system with an evolutionary computation algorithm. *Energy*, 36(5), 2440–2449.

Leaf Community, Loccioni Energy (n.d.). Leaf Community [Online]. Available: https://www.loccioni.com/en/waves/leaf-community/.

Lee, Y.M., Horesh, R., Liberti, L. (2015). Simulation and optimization of energy efficient operation of HVAC system as demand response with distributed energy resources. *2015 Winter Simulation Conference*, Huntington Beach, 991–999.

Li, Y., Yang, Z., Li, G., Mu, Y., Zhao, D., Chen, C., Shen, B. (2018). Optimal scheduling of isolated microgrid with an electric vehicle battery swapping station in multi-stakeholder scenarios: A bi-level programming approach via real-time pricing. *Applied Energy*, 232, 54–68.

Li, Y., Yang, Z., Zhao, D., Lei, H., Cui, B., Li, S. (2019). Incorporating energy storage and user experience in isolated microgrid dispatch using a multi-objective model. *IET Renewable Power Generation*, 13(6), 973–981.

Lin, H.-W., Hong, T. (2013). On variations of space-heating energy use in office buildings. *Applied Energy*, 111, 515–528.

Loccioni Group (2017). Loccioni [Online]. Available: https://www.loccioni.com/en/ [accessed February 13, 2017].

Losi, A., Mancarella, P., Vicino, A. (eds) (2015). *Integration of Demand Response into the Electricity Chain*. ISTE Ltd, London, and John Wiley & Sons, New York.

Lu, N., Du, P., Guo, X., Greitzer, F.L. (2012). Smart meter data analysis. *Proceedings of the Sixth IEEE PES Transmission and Distribution 2012*, Orlando, 1–6.

Ma, O., Alkadi, N., Cappers, P., Denholm, P., Dudley, J., Goli, S., Hummon, M., Kiliccote, S., MacDonald, J., Matson, N., Olsen, D., Rose, C., Sohn, M.D., Starke, M., Kirby, B., O'Malley, M. (2013). Demand response for ancillary services. *IEEE Transactions on Smart Grid*, 4(4), 1988–1995.

Ma, K., Yuan, C., Yang, J., Liu, Z., Guan, X. (2017). Switched control strategies of aggregated commercial HVAC systems for demand response in smart grids. *Energies*, 10(7), 953.

Marinakis, V., Doukas, H. (2018). An advanced IoT-based system for intelligent energy management in buildings. *Sensors*, 18, 610.

Marszal, A.J., Heiselberg, P., Bourrelle, J.S., Musall, E., Voss, K., Sartori, I., Napolitano, A. (2011). Zero energy building: A review of definitions and calculation methodologies. *Energy and Buildings*, 43(4), 971–979.

Mathieu, J.L. (2012). Modeling, analysis, and control of demand response resources. PhD Thesis, University of California, Berkeley.

Mavrigiannaki, A., Kampelis, N., Kolokotsa, D., Marchegiani, D., Standardi, L., Isidori, D., Christalli, C. (2017). Development and testing of a micro-grid excess power production forecasting algorithms. *Energy Procedia*, 134, 654–663.

McCollum, D., Gomez, L., Riahi, E.K., Parkinson, S. (2017). A guide to SDG interactions: From science to implementation [Online]. Available: http://www.icsu.org/publications/a-guide-to-sdg-interactions-from-science-to-implementation.

Menezes, A.C., Cripps, A., Bouchlaghem, D., Buswell, R. (2012). Predicted vs. actual energy performance of non-domestic buildings: Using post-occupancy evaluation data to reduce the performance gap. *Applied Energy*, 97, 355–364.

Mohseni, A., Mortazavi, S.S., Ghasemi, A., Nahavandi, A., Talaei abdi, M. (2017). The application of household appliances' flexibility by set of sequential uninterruptible energy phases model in the day-ahead planning of a residential microgrid. *Energy*, 139, 315–328.

Motegi, N., Piette, M., Watson, D. (2007). Introduction to commercial building control strategies and techniques for demand response. LBNL Report 59975, Berkeley, California.

Murty, V.V.S.N., Kumar, A. (2020). Multi-objective energy management in microgrids with hybrid energy sources and battery energy storage systems. *Protection and Control of Modern Power Systems*, 5, 2.

MyLeaf Platform–Loccioni Group (2018). MyLeaf [Online]. Available: https://myleaf2.loccioni.com [accessed July 26, 2018].

NERC (2011). Demand response availability data system (DADS): Phase I & II final report. North American Electric Reliability Corporation, Atlanta.

Nizami, M.S.H., Hossain, M.J., Amin, B.M.R., Fernandez, E. (2020). A residential energy management system with bi-level optimization-based bidding strategy for day-ahead bi-directional electricity trading. *Applied Energy*, 261, 114322.

Okochi, G.S., Yao, Y. (2016). A review of recent developments and technological advancements of variable-air-volume (VAV) air-conditioning systems. *Renewable and Sustainable Energy Reviews*, 59, 784–817.

OpenADR Alliance (2014). OpenADR 2.0 demand response program implementation guide. OpenADR Alliance.

OpenStudio (2017). OpenStudio [Online]. Available: https://www.openstudio.net [accessed February 13, 2017].

Ou, T.-C. (2012). A novel unsymmetrical faults analysis for microgrid distribution systems. *International Journal of Electrical Power & Energy Systems*, 43(1), 1017–1024.

Ou, T.-C. (2013). Ground fault current analysis with a direct building algorithm for microgrid distribution. *International Journal of Electrical Power & Energy Systems*, 53, 867–875.

Ou, T.-C., Hong, C.-M. (2014). Dynamic operation and control of microgrid hybrid power systems. *Energy*, 66, 314–323.

Ou, T.-C., Lu, K.-H., Huang, C.-J. (2017). Improvement of transient stability in a hybrid power multi-system using a designed NIDC (Novel Intelligent Damping Controller). *Energies*, 10(4), 488.

Palensky, P., Dietrich, D. (2011). Demand side management: Demand response, intelligent energy systems, and smart loads. *IEEE Transactions on Industrial Informatics*, 7(3), 381–388.

Papantoniou, S., Kolokotsa, D., Kalaitzakis, K. (2015). Building optimization and control algorithms implemented in existing BEMS using a web based energy management and control system. *Energy and Buildings*, 98, 45–55.

Pérez-Lombard, L., Ortiz, J., Pout, C. (2008). A review on buildings energy consumption information. *Energy and Buildings*, 40(3), 394–398.

Pombeiro, H., Machado, M.J., Silva, C. (2017). Dynamic programming and genetic algorithms to control an HVAC system: Maximizing thermal comfort and minimizing cost with PV production and storage. *Sustainable Cities and Society*, 34, 228–238.

Provata, E., Kolokotsa, D., Papantoniou, S., Pietrini, M., Giovannelli, A., Romiti, G. (2015). Development of optimization algorithms for the Leaf Community microgrid. *Renewable Energy*, 74, 782–795.

Raza, M.Q., Khosravi, A. (2015). A review on artificial intelligence based load demand forecasting techniques for smart grid and buildings. *Renewable and Sustainable Energy Reviews*, 50, 1352–1372.

Rietveld, G., Clarkson, P., Wright, P., Pogliano, U., Braun, J., Kokalj, M., Lapuh, R., Zisky, N. (2012). Measurement infrastructure for observing and controlling smart electrical grids. *3rd IEEE PES Innovative Smart Grid Technologies Europe*, Berlin, 1–8.

Risbeck, M.J., Maravelias, C.T., Rawlings, J.B., Turney, R.D. (2017). A mixed-integer linear programming model for real-time cost optimization of building heating, ventilation, and air conditioning equipment. *Energy and Buildings*, 142, 220–235.

Safamehr, H., Rahimi-Kian, A. (2015). A cost-efficient and reliable energy management of a micro-grid using intelligent demand-response program. *Energy*, 91, 283–293.

Satre-Meloy, A., Diakonova, M., Grünewald, P. (2020). Cluster analysis and prediction of residential peak demand profiles using occupant activity data. *Applied Energy*, 260, 114246.

SEDC (2015). Mapping demand response in Europe Today 2015. Report, SEDC, Brussels.

Serra, J., Pubill, D., Vázquez, M.Á., Verikoukis, C. (2014). Experimental evaluation of an HVAC system under dynamic pricing with comfort constraints. *Innovative Smart Grid Technologies*, Washington, DC, 1–5.

Shabani, M.J., Moghaddas-Tafreshi, S.M. (2020). Fully-decentralized coordination for simultaneous hydrogen, power, and heat interaction in a multi-carrier-energy system considering private ownership. *Electric Power Systems Research*, 180, 106099.

Shahidehpour, M., Yamin, H., Li, Z. (2002). *Market Operations in Electric Power Systems: Forecasting, Scheduling, and Risk Management*. Wiley – IEEE, New York.

Shakeri, M., Shayestegan, M., Abunima, H., Reza, S.M.S., Akhtaruzzaman, M., Alamoud, A.R.M., Sopian, K., Amin, N. (2017). An intelligent system architecture in home energy management systems (HEMS) for efficient demand response in smart grid. *Energy and Buildings*, 138, 154–164.

Shariatzadeh, F., Mandal, P., Srivastava, A.K. (2015). Demand response for sustainable energy systems: A review, application and implementation strategy. *Renewable and Sustainable Energy Reviews*, 45, 343–350.

Shinkhede, S., Joshi, S.K. (2014). Hybrid optimization algorithm for distributed energy resource (DER) management. *Innovative Applications of Computational Intelligence on Power, Energy and Controls with their Impact on Humanity*, Ghaziabad, 411–416.

Siano, P. (2014). Demand response and smart grids – A survey. *Renewable and Sustainable Energy Reviews*, 30, 461–478.

SketchUp (2017). 3D modeling for everyone, SketchUp [Online]. Available: http://www.sketchup.com/ [accessed February 13, 2017].

Song, H.Y., Lee, G.S., Yoon, Y.T. (2019). Optimal operation of critical peak pricing for an energy retailer considering balancing costs. *Energies*, 12, 4658.

Susowake, Y., Masrur, H., Yabiku, T., Senjyu, T., Howlader, A.M., Abdel-Akher, M., Hemeida, A.M. (2019). A multi-objective optimization approach towards a proposed smart apartment with demand-response in Japan. *Energies*, 13(1), 127.

Torriti, J., Hassan, M.G., Leach, M. (2010). Demand response experience in Europe: Policies, programmes and implementation. *Energy*, 35(4), 1575–1583.

Triki, C., Violi, A. (2009). Dynamic pricing of electricity in retail markets. *4or*, 7(1), 21–36.

Tsekouras, G.J., Kanellos, F.D., Mastorakis, N. (2015). Short term load forecasting in electric power systems with artificial neural networks. In *Computational Problems in Science and Engineering*, Mastorakis, N., Bulucea, A., Tsekouras, G. (eds). Springer, Cham, 343.

Tsitsiklis, J.N., Xu, Y. (2015). Pricing of fluctuations in electricity markets. *European Journal of Operational Research*, 246, 199–208.

Tsui, K.M., Chan, S.C. (2012). Demand response optimization for smart home scheduling under real-time pricing. *IEEE Transactions on Smart Grid*, 3(4), 1812–1821.

U.S. Department of Energy (2006). Benefits of demand response in electricity markets and recommendations for achieving them. A report to the United States Congress Pursuant to Section 1252 of the Energy Policy Act of 2005. Washington, DC.

U.S. Department of Energy (2015). EnergyPlus energy simulation software. Energy Efficiency and Renewable Energy. Building Technologies Office [Online]. Available: http://apps1.eere.energy.gov/buildings/energyplus/ [accessed February 13, 2017].

U.S. Department of Energy (2019). EnergyPlus™ version 9.1.0 documentation. Engineering reference, U.S. Department of Energy.

Vallés, M., Reneses, J., Cossent, R., Frías, P. (2016). Regulatory and market barriers to the realization of demand response in electricity distribution networks: A European perspective. *Electric Power Systems Research*, 140, 689–698.

Vardakas, J., Zorba, N., Verikoukis, C. (2015). A survey on demand response in smart grids. *IEEE Transactions on Industrial Informatics*, 11(3), 1–8.

Wang, S., Ma, Z. (2008). Supervisory and optimal control of building HVAC systems: A review. *HVAC&R Research*, 14(1), 3–32.

Wang, Y., Huang, Y., Wang, Y., Zeng, M., Li, F., Wang, Y., Zhang, Y. (2018). Energy management of smart micro-grid with response loads and distributed generation considering demand response. *Journal of Cleaner Production*, 197, 1069–1083.

WeatherSpark (n.d.) Average weather for Ancona, Italy – WeatherSpark [Online]. Available: https://weatherspark.com/averages/32287/Ancona-Marche-Italy/ [accessed March 6, 2017].

Webster, L., Bradford, J., Sartor, D., Shonder, J., Atkin, E., Dunnivant, S., Frank, D., Franconi, E., Jump, D., Schiller, S., Stetz, M., Slattery, B. (2015). M&V guidelines: Measurement and verification for performance-based contracts version 4.0. Guidelines, US DoE.

Yang, J., Tian, Z., Ma, K. (2018). A demand-side pricing strategy based on Bayesian game. *15th International Conference on Control, Automation, Robotics and Vision*, Singapore, 146–150.

Yin, R., Kara, E.C., Li, Y., DeForest, N., Wang, K., Yong, T., Stadler, M. (2016). Quantifying flexibility of commercial and residential loads for demand response using setpoint changes. *Applied Energy*, 177, 149–164.

Yuan, D., Lu, Z., Zhang, J., Li, X. (2019). A hybrid prediction-based microgrid energy management strategy considering demand-side response and data interruption. *International Journal of Electrical Power & Energy Systems*, 113, 139–153.

Yuce, B., Mourshed, M., Rezgui, Y. (2017). An ANN-based energy forecasting framework for the district level smart grids. In *Smart Grid Inspired Future Technologies*, Hu, J., Leung, V.C.M., Yang, K., Zhang, Y., Gao, J., Yang, S. (eds). Springer International Publishing, New York.

Zangheri, P., Pagliano, L., Bürger, V. (2012). The challenges, dynamics and activities in the building sector and its energy demand in Italy. D2.1 of WP2 from Entranze Project, Entranze.

Zhang, C., Ding, Y., Nordentoft, N.C., Pinson, P., Østergaard, J. (2014a). FLECH: A Danish market solution for DSO congestion management through DER flexibility services. *Journal of Modern Power Systems and Clean Energy*, 2(2), 126–133.

Zhang, Y., Zeng, P., Zang, C. (2014b). Multi-objective optimal control algorithm for HVAC based on particle swarm optimization. *Fifth International Conference on Intelligent Control and Information Processing*, Dalian, 417–423.

Zhou, B., Li, W., Chan, K.W., Cao, Y., Kuang, Y., Liu, X., Wang, X. (2016). Smart home energy management systems: Concept, configurations, and scheduling strategies. *Renewable and Sustainable Energy Reviews*, 61, 30–40.

Zia, M.F., Elbouchikhi, E., Benbouzid, M. (2018). Microgrids energy management systems: A critical review on methods, solutions, and prospects. *Applied Energy*, 222, 1033–1055.

List of Authors

Cristina CRISTALLI
Research for Innovation
AEA s.r.l.
Angeli di Rosora
Italy

Marina Kyprianou DRACOU
The Cyprus Institute
Nicosia
Cyprus

Konstantinos GOBAKIS
Energy Management
in the Built Environment
Research Lab
Environmental
Engineering School
Technical University of Crete
Chania
Greece

Daniela ISIDORI
Research for Innovation
AEA s.r.l.
Angeli di Rosora
Italy

Konstantinos KALAITZAKIS
Circuits, Sensors and
Renewable Energy Sources
Laboratory
Technical University of Crete
Chania
Greece

Nikos KAMPELIS
Energy Management
in the Built Environment
Research Lab
Environmental Engineering
School
Technical University of Crete
Chania
Greece

Theoni KARLESSI
Physics Department
Section of Environmental
Physics and Meteorology
Group Building
Environmental Studies
University of Athens
Greece

Denia KOLOKOTSA
Energy Management
in the Built Environment
Research Lab
Environmental Engineering
School
Technical University of Crete
Chania
Greece

Fabio Maria MONTAGNINO
IDEA s.r.l.
Termini Imerese
Italy

Alaric MONTENON
The Cyprus Institute
Nicosia
Cyprus

Pietro MURATORE
IDEA s.r.l.
Termini Imerese
Italy

Filippo PAREDES
IDEA s.r.l.
Termini Imerese
and
Consorzio ARCA
Viale delle Scienze
Palermo
Italy

Andri PYRGOU
The Cyprus Institute
Nicosia
Cyprus

Mattheos SANTAMOURIS
School of Built Environment
University of New South
Wales
Sydney
Australia

Nikolaos SIFAKIS
Renewable and Sustainable
Energy Systems Laboratory
Environmental
Engineering School
Technical University of Crete
Chania
Greece

Laura STANDARDI
Research for Innovation
AEA s.r.l.
Angeli di Rosora
Italy

Elisavet TSEKERI
Energy Management
in the Built Environment
Research Lab
Environmental Engineering
School
Technical University of Crete
Chania
Greece

Vagias VAGIAS
Energy Management
in the Built Environment
Research Lab
Environmental Engineering
School
Technical University of Crete
Chania
Greece

Luca VENEZIA
IDEA s.r.l.
Termini Imerese
Italy

Index

E, F

G

H, I

L, M

V

W, Z

Other titles from

ISTE

in

Science, Society and New Technologies

2020

BARNOUIN Jacques
The World's Construction Mechanism: Trajectories, Imbalances and the Future of Societies
(Interdisciplinarity between Biological Sciences and Social Sciences Set – Volume 4)

ÇAĞLAR Nur, CURULLI Irene G., SIPAHIOĞLU Işıl Ruhi, MAVROMATIDIS Lazaros
Thresholds in Architectural Education (Engineering, Energy and Architecture Set – Volume 7)

ETCHEVERRIA Olivier
The Restaurant, A Geographical Approach: From Invention to Gourmet Tourist Destinations
(Tourism and Mobility Systems Set – Volume 3)

JEANNERET Yves
The Trace Factory
(Traces Set – Volume 3)

KATSAFADOS Petros, MAVROMATIDIS Elias, SPYROU Christos
Numerical Weather Prediction and Data Assimilation (Engineering, Energy and Architecture Set – Volume 6)

MARTI Caroline
Cultural Mediations of Brands: Unadvertization and Quest for Authority
(Communication Approaches to Commercial Mediation Set – Volume 1)

MAVROMATIDIS Lazaros E.
Climatic Heterotopias as Spaces of Inclusion: Sew Up the Urban Fabric
(Research in Architectural Education Set – Volume 1)

SCHMITT Daniel, THÉBAULT Marine, BURCZYKOWSKI Ludovic
Image Beyond the Screen: Projection Mapping

VIOLIER Philippe, with the collaboration of TAUNAY Benjamin
The Tourist Places of the World
(Tourism and Mobility Systems Set – Volume 2)

2019

BRIANÇON Muriel
The Meaning of Otherness in Education: Stakes, Forms, Process, Thoughts and Transfers
(Education Set – Volume 3)

DESCHAMPS Jacqueline
Mediation: A Concept for Information and Communication Sciences
(Concepts to Conceive 21st Century Society Set – Volume 1)

DOUSSET Laurent, PARK Sejin, GUILLE-ESCURET Georges
Kinship, Ecology and History: Renewal of Conjunctures
(Interdisciplinarity between Biological Sciences and Social Sciences Set – Volume 3)

DUPONT Olivier
Power
(Concepts to Conceive 21st Century Society Set – Volume 2)

FERRARATO Coline
Prospective Philosophy of Software: A Simondonian Study

GUAAYBESS Tourya
The Media in Arab Countries: From Development Theories to Cooperation Policies

HAGÈGE Hélène
Education for Responsibility
(Education Set – Volume 4)

LARDELLIER Pascal
The Ritual Institution of Society
(Traces Set – Volume 2)

LARROCHE Valérie
The Dispositif
(Concepts to Conceive 21st Century Society Set – Volume 3)

LATERRASSE Jean
Transport and Town Planning: The City in Search of Sustainable Development

LENOIR Virgil Cristian
Ethically Structured Processes
(Innovation and Responsibility Set – Volume 4)

LOPEZ Fanny, PELLEGRINO Margot, COUTARD Olivier
Local Energy Autonomy: Spaces, Scales, Politics
(Urban Engineering Set – Volume 1)

METZGER Jean-Paul
Discourse: A Concept for Information and Communication Sciences
(Concepts to Conceive 21st Century Society Set – Volume 4)

MICHA Irini, VAIOU Dina
Alternative Takes to the City
(Engineering, Energy and Architecture Set – Volume 5)

PÉLISSIER Chrysta
Learner Support in Online Learning Environments

PIETTE Albert
Theoretical Anthropology or How to Observe a Human Being
(Research, Innovative Theories and Methods in SSH Set – Volume 1)

PIRIOU Jérôme
The Tourist Region: A Co-Construction of Tourism Stakeholders
(Tourism and Mobility Systems Set – Volume 1)

PUMAIN Denise
Geographical Modeling: Cities and Territories
(Modeling Methodologies in Social Sciences Set – Volume 2)

WALDECK Roger
Methods and Interdisciplinarity
(Modeling Methodologies in Social Sciences Set – Volume 1)

2018

BARTHES Angela, CHAMPOLLION Pierre, ALPE Yves
Evolutions of the Complex Relationship Between Education and Territories
(Education Set – Volume 1)

BÉRANGER Jérôme
The Algorithmic Code of Ethics: Ethics at the Bedside of the Digital Revolution
(Technological Prospects and Social Applications Set – Volume 2)

DUGUÉ Bernard
Time, Emergences and Communications
(Engineering, Energy and Architecture Set – Volume 4)

GEORGANTOPOULOU Christina G., GEORGANTOPOULOS George A.
Fluid Mechanics in Channel, Pipe and Aerodynamic Design Geometries 1
(Engineering, Energy and Architecture Set – Volume 2)

GEORGANTOPOULOU Christina G., GEORGANTOPOULOS George A.
Fluid Mechanics in Channel, Pipe and Aerodynamic Design Geometries 2
(Engineering, Energy and Architecture Set – Volume 3)

GUILLE-ESCURET Georges
Social Structures and Natural Systems: Is a Scientific Assemblage Workable?
(Social Interdisciplinarity Set – Volume 2)

LARINI Michel, BARTHES Angela
Quantitative and Statistical Data in Education: From Data Collection to Data Processing
(Education Set – Volume 2)

LELEU-MERVIEL Sylvie
Informational Tracking
(Traces Set – Volume 1)

SALGUES Bruno
Society 5.0: Industry of the Future, Technologies, Methods and Tools
(Technological Prospects and Social Applications Set – Volume 1)

TRESTINI Marc
Modeling of Next Generation Digital Learning Environments: Complex Systems Theory

2017

ANICHINI Giulia, CARRARO Flavia, GESLIN Philippe, GUILLE-ESCURET Georges
Technicity vs Scientificity – Complementarities and Rivalries
(Interdisciplinarity between Biological Sciences and Social Sciences Set – Volume 2)

DUGUÉ Bernard
Information and the World Stage – From Philosophy to Science, the World of Forms and Communications
(Engineering, Energy and Architecture Set – Volume 1)

GESLIN Philippe
Inside Anthropotechnology – User and Culture Centered Experience
(Social Interdisciplinarity Set – Volume 1)

GORIA Stéphane
Methods and Tools for Creative Competitive Intelligence

KEMBELLEC Gérald, BROUDOUS EVELYNE
Reading and Writing Knowledge in Scientific Communities: Digital Humanities and Knowledge Construction

MAESSCHALCK Marc
Reflexive Governance for Research and Innovative Knowledge (Responsible Research and Innovation Set - Volume 6)

PARK Sejin, GUILLE-ESCURET Georges
Sociobiology vs Socioecology: Consequences of an Unraveling Debate (Interdisciplinarity between Biological Sciences and Social Sciences Set – Volume 1)

PELLÉ Sophie
Business, Innovation and Responsibility (Responsible Research and Innovation Set – Volume 7)

2016

BRONNER Gérald
Belief and Misbelief Asymmetry on the Internet

EL FALLAH SEGHROUCHNI Amal, ISHIKAWA Fuyuki, HÉRAULT Laurent, TOKUDA Hideyuki
Enablers for Smart Cities

GIANNI Robert
Responsibility and Freedom (Responsible Research and Innovation Set – Volume 2)

GRUNWALD Armin
The Hermeneutic Side of Responsible Research and Innovation (Responsible Research and Innovation Set – Volume 5)

LAGRAÑA Fernando
E-mail and Behavioral Changes: Uses and Misuses of Electronic Communications

LENOIR Virgil Cristian
Ethical Efficiency: Responsibility and Contingency (Responsible Research and Innovation Set – Volume 1)

MAESSCHALCK Marc
Reflexive Governance for Research and Innovative Knowledge
(Responsible Research and Innovation Set – Volume 6)

PELLÉ Sophie, REBER Bernard
From Ethical Review to Responsible Research and Innovation
(Responsible Research and Innovation Set – Volume 3)

REBER Bernard
Precautionary Principle, Pluralism and Deliberation: Sciences and Ethics
(Responsible Research and Innovation Set – Volume 4)

VENTRE Daniel
Information Warfare – 2nd edition

www.ingramcontent.com/pod-product-compliance
Lightning Source LLC
LaVergne TN
LVHW011803200926
848484LV00003B/770